EAA Series–Textbook

EAA series is successor of the EAA Lecture Notes and supported by the European Actuarial Academy (EAA GmbH), founded on the 29 August, 2005 in Cologne (Germany) by the Actuarial Associations of Austria, Germany, the Netherlands and Switzerland. EAA offers actuarial education including examination, permanent education for certified actuaries and consulting on actuarial education.

actuarial-academy.com

For further titles published in this series, please go to
http://www.springer.com/series/7879

Francesca Biagini • Andreas Richter •
Harris Schlesinger
Editors

Risk Measures and Attitudes

Springer

Editors
Francesca Biagini
Department of Mathematics
University of Munich
Munich, Germany

Harris Schlesinger
University of Alabama
Tuscaloosa, AL, USA

Andreas Richter
Institute for Risk Management and Insurance
University of Munich
Munich, Germany

ISSN 1869-6929 ISSN 1869-6937 (electronic)
EAA Series
ISBN 978-1-4471-4925-5 ISBN 978-1-4471-4926-2 (eBook)
DOI 10.1007/978-1-4471-4926-2
Springer London Heidelberg New York Dordrecht

Library of Congress Control Number: 2013931507

AMS Subject Classification: 00B20, 60G46, 60H05

Printed on acid-free paper

Springer is part of Springer Science+Business Media (www.springer.com)

Preface

In ancient times, risk was viewed simply as the will of the gods. And tempting fate was only a way that man could anger the gods (Bernstein 1996). But in the seventeenth century, Blaise Pascal and Pierre Fermat set the mathematical foundations for modern-day probability theory. It was not until the following century that another mathematician, Daniel Bernoulli (1738), pointed out how decisions made in the face of risk were not typically based only on the expected outcomes. Indeed, Bernoulli illustrated how a diminishing marginal valuation of additional wealth could explain an aversion to risk taking. Bernoulli's insight is still relevant today with regards to risk aversion, which is the most basic of all risk attitudes. Similarly, the second moment of any risky distribution of monetary payoffs was relevant as one measure of risk in decision making.

Only more recently did we realize that characteristics of risky distributions rely on more than just its moments. Likewise, attitudes towards taking a given risk depend on more than simple risk aversion. Stochastic dominance and other strong partial orderings of risky distributions led the way to developing stronger risk measures, which could then be used to choose among various risky alternatives. Likewise, attitudes towards higher orders of risk can play a crucial role in analyzing decisions made among risky choices.

The related topics of risk measures and risk attitudes were the focus of a small conference held at the Ludwig-Maximilians University in Munich in December 2010. Several of the papers either presented at that conference or generated by discussions during the meetings are included in this Symposium volume.

In the first chapter, Patrick Cheridito, Samuel Drapeau and Michael Kupper establish a quasi-convexity duality setting for comparing risky distributions (lotteries) that have a compact support. The authors introduce specific types of lower semicontinuity that allow for a convenient functional form in these comparisons and for a robust representation of risk preferences on lotteries with compact support. As a useful illustration, the authors model "Value at Risk" as a functional on a space of lotteries.

Michel Denuit, Louis Eeckhoudt, Ilia Tsetlin and Robert Winkler examine next multivariate stochastic dominance as a tool for partially ordering risky alternatives.

The authors provide definitions of multivariate risk averse and multivariate risk seeking based on stochastic dominance relationships. These definitions are used to reveal some interesting properties of additive or multiplicative background risks. The approach taken here is compared to several other stochastic orders that appear in the literature.

In the third chapter, Jörn Dunkel and Stefan Weber consider some issues in looking at the downside risk associated with potential default in credit markets. They show how the current industry standard Value-at-Risk models do not adequately measure the level of risk and how the introduction of well-defined, tail-sensitive shortfall risk measures (SR) can dramatically improve both the management and the regulation of credit risk. In particular, the authors introduce a novel Monte Carlo approach for the efficient computation of SR by combining stochastic root-approximation algorithms with variance reduction techniques.

Finally, Claudio Fontana and Wolfgang Runggaldier examine a class of Itô-process models for investment markets for which local martingale measures might not exist. In this setting they discuss several notions of no-arbitrage and discuss several sufficient and necessary conditions for their validity in terms of the integrability of the market price of risk process and of the existence of martingale deflators. This is connected to the Growth-Optimal-Portfolio (GOP), which can be explicitly characterized in a unique way and possesses the numéraire property (i.e. all admissible processes when denominated in units of the numéraire are supermartingales). Another major issue of this chapter is the valuation and hedging of contingent claims. In particular, the authors show that financial markets may be viable and complete without the existence of a martingale measure. Contingent claims can be then evaluated by using for example real-world pricing, upper-hedging pricing or utility indifference valuation. In the case of a complete financial market model, these three methods deliver the same valuation formula, given by the GOP-discounted expected value under the original (real-world) probability measure. Some of the results presented in this chapter are already well known. However the authors add also in this case new interesting contributions to the established theory by providing simple and transparent proofs by exploiting the Itô-process structure of the underlying model.

Overall, the papers presented in this volume show how modern theory now incorporates newer approaches to both risk measures and to risk attitudes. They also provide useful illustrations of how these two concepts are inevitably intertwined. Over the coming year, the integrative nature of these concepts will likely become even more transparent. We hope that the reader will find the topics included in this Symposium volume of interest; and we hope that this interest translates into further journeys into this fertile area of research.

The editors would like to thank Irena Grgic for his assistance in compiling this volume and all the authors for their contributions.

Munich, Germany Francesca Biagini
Andreas Richter
Tuscaloosa, USA Harris Schlesinger

Contents

Contributors

Patrick Cheridito Princeton University, Princeton, NJ, USA

Michel Denuit Institut des Sciences Actuarielles & Institut de Statistique, Université Catholique de Louvain, Louvain-la-Neuve, Belgium

Samuel Drapeau Humboldt University Berlin, Berlin, Germany

Jörn Dunkel Department of Applied Mathematics and Theoretical Physics, University of Cambridge, Cambridge, UK

Louis Eeckhoudt IESEG School of Management, LEM, Université Catholique de Lille, Lille, France; CORE, Université Catholique de Louvain, Louvain-la-Neuve, Belgium

Claudio Fontana INRIA Paris-Rocquencourt, Le Chesnay Cedex, France

Michael Kupper Humboldt University Berlin, Berlin, Germany

Wolfgang J. Runggaldier Department of Mathematics, University of Padova, Padova, Italy

Ilia Tsetlin INSEAD, Singapore, Singapore

Stefan Weber Institut für Mathematische Stochastik, Leibniz Universität Hannover, Hannover, Germany

Robert L. Winkler Fuqua School of Business, Duke University, Durham, NC, USA

Part I
Risk Attitudes

Chapter 1
Weak Closedness of Monotone Sets of Lotteries and Robust Representation of Risk Preferences

Patrick Cheridito, Samuel Drapeau, and Michael Kupper

Keywords Risk preferences · Robust representations · Lotteries with compact support · Monotonicity

1.1 Introduction

We consider a risk preference given by a total preorder $\succcurlyeq$ on the set $\mathcal{M}_{1,c}$ of probability distributions on $\mathbb{R}$ with compact support, that is, a transitive binary relation such that for all $\mu, \nu \in \mathcal{M}_{1,c}$, one has $\mu \succcurlyeq \nu$ or $\mu \preccurlyeq \nu$ or both. Elements μ of $\mathcal{M}_{1,c}$ are understood as lotteries, and $\mu \succcurlyeq \nu$ means that μ is at least as risky as ν.

The goal of the paper is to provide conditions under which $\succcurlyeq$ has a numerical representation of the form

$$\rho(\mu) = \sup_{l \in L} R\big(l, \langle l, \mu \rangle\big), \tag{1.1}$$

where L is the set of all nonincreasing continuous functions $l : \mathbb{R} \to \mathbb{R}$, $\langle l, \mu \rangle := \int_{\mathbb{R}} l \, d\mu$, and $R : L \times \mathbb{R} \to [-\infty, +\infty]$ is a function satisfying

(R1) $R(l, s)$ is left-continuous and nondecreasing in s;
(R2) R is quasi-concave in (l, s);
(R3) $R(\lambda l, s) = R(l, s/\lambda)$ for all $l \in L$, $s \in \mathbb{R}$ and $\lambda > 0$;
(R4) $\inf_{s \in \mathbb{R}} R(l^1, s) = \inf_{s \in \mathbb{R}} R(l^2, s)$ for all $l^1, l^2 \in L$;
(R5) $R^+(l, s) := \inf_{t > s} R(l, t)$ is upper semi-continuous in l with respect to $\sigma(C, \mathcal{M}_{1,c})$, where C denotes the space of all continuous functions $f : \mathbb{R} \to \mathbb{R}$.

P. Cheridito
Princeton University, Princeton, NJ 08544, USA
e-mail: dito@princeton.edu

S. Drapeau (✉) · M. Kupper
Humboldt University Berlin, Unter den Linden 6, 10099 Berlin, Germany
e-mail: drapeau@mathematik.hu-berlin.de

M. Kupper
e-mail: kupper@mathematik.hu-berlin.de

F. Biagini et al. (eds.), *Risk Measures and Attitudes*, EAA Series,
DOI 10.1007/978-1-4471-4926-2_1, © Springer-Verlag London 2013

Relation (1.1) can be viewed as a robust representation of risk. Each $l \in L$ induces a risk order on $\mathcal{M}_{1,c}$ through the affine mapping $\mu \mapsto \langle l, \mu \rangle$. Relation (1.1) takes all these orders into account but gives them different impacts by weighing them according to the risk function R. It follows from (R1) that every mapping $\rho : \mathcal{M}_{1,c} \to [-\infty, \infty]$ of the form (1.1) has the following three properties:

(A1) quasi-convexity;
(A2) monotonicity with respect to first-order stochastic dominance;
(A3) lower semicontinuity with respect to the weak topology $\sigma(\mathcal{M}_{1,c}, C)$.

Sufficient conditions for preferences on lotteries to have affine representations go back to von Neumann and Morgenstern (1947). For an overview of subsequent extensions, we refer to Fishburn (1982). Representations of the form (1.1) have recently been given by Cerreia-Vioglio (2009) and Drapeau and Kupper (2010). The contribution of this paper is that we do not make assumptions on $\succcurlyeq$ involving the topology $\sigma(\mathcal{M}_{1,c}, C)$ since they are technical and difficult to check empirically. Instead, we provide conditions with a certain normative appeal and show that they imply that the sublevel sets of $\succcurlyeq$ are closed in $\sigma(\mathcal{M}_{1,c}, C)$. Similar results are given in Delbaen et al. (2011) for preferences satisfying the independence and Archimedean axioms. For automatic continuity and representation results on risk measures defined on spaces of random variables, we refer to Cheridito and Li (2008, 2009) and the references therein.

As an example, we discuss Value-at-Risk. It is well known that as a function of random variables, it is not quasi-convex. But Value-at-Risk only depends on the distribution μ_X of a random variable X, and convex combinations act differently on distributions than on random variables. Except for trivial cases, one has $\lambda\mu_X + (1-\lambda)\mu_Y \neq \mu_{\lambda X+(1-\lambda)Y}$. It turns out that as a function on $\mathcal{M}_{1,c}$, Value-at-risk is quasi-convex, $\sigma(\mathcal{M}_{1,c}, C)$-lower semicontinuous and monotone with respect to first-order stochastic dominance. As a consequence, it can be represented in the form (1.1); see Example 1.2.4 below.

The rest of the paper is organized as follows. In Sect. 1.2 we introduce the conditions we need to show that $\succcurlyeq$ has a representation of the form (1.1) and state the main results, Theorems 1.2.1 and 1.2.2. Section 1.3 contains a discussion of the weak topology $\sigma(\mathcal{M}_{1,c}, C)$ and the proof of Theorem 1.2.1.

1.2 Robust Representation of Risk Preferences on Lotteries

To formulate the conditions (C1)–(C3) below, we need the following notation:

- By $\mathcal{M}_1$ we denote the set of all probability distributions on $\mathbb{R}$. For $\mu \in \mathcal{M}_1$, we set $F_\mu(x) := \mu(-\infty, x]$ and

$$\mu_* := \sup\{x \in \mathbb{R} : F_\mu(x) = 0\} \quad \text{and} \quad \mu^* := \inf\{x \in \mathbb{R} : F_\mu(x) = 1\},$$

where $\sup \emptyset := -\infty$ and $\inf \emptyset := +\infty$.

- By $\trianglerighteq$ we denote first-order stochastic dominance on $\mathcal{M}_1$, that is,

$$\mu \trianglerighteq \nu \quad :\Leftrightarrow \quad F_\mu(x) \le F_\nu(x) \quad \text{for all } x \in \mathbb{R}.$$

- For $m \in \mathbb{R}$ and $\mu \in \mathcal{M}_1$, we denote by $T_m\mu$ the shifted distribution given by $T_m\mu(A) = \mu(A - m)$.

To show that the risk preference $\succcurlyeq$ has a representation of the form (1.1), we assume that for each $\nu \in \mathcal{M}_{1,c}$, the sublevel set

$$\mathcal{S}_\nu := \{\mu \in \mathcal{M}_{1,c} : \mu \preccurlyeq \nu\}$$

satisfies the following conditions:

(C1) $\mathcal{S}_\nu$ is convex;
(C2) If $T_m\mu \in \mathcal{S}_\nu$ for all $m > 0$, then $\mu \in \mathcal{S}_\nu$;
(C3) If $\mu \in \mathcal{M}_{1,c}$ has the property that for every $\lambda \in [0,1)$ and each $\eta \in \mathcal{M}_1$ with $\eta_* \ge \mu^*$ and $\eta^* = +\infty$, one has $\lambda\mu + (1-\lambda)\eta \trianglerighteq \tau$ for some $\tau \in \mathcal{S}_\nu$, then $\mu \in S_\nu$.

First, let us note that (C3) implies

(C4) $\mu \preccurlyeq \nu$ whenever $\mu \trianglerighteq \nu$,

which is a standard assumption. It just means that "more is better" or "more is less risky." Assumption (C1) is also standard and corresponds to the idea that "averages are better than extremes"or "diversification should not increase the risk." As for (C2) and (C3), they allow us to deduce that all sublevel sets $\mathcal{S}_\nu$ are closed in $\sigma(\mathcal{M}_{1,c}, C)$, which is needed to derive a representation of the form (1.1). But $\sigma(\mathcal{M}_{1,c}, C)$-closedness is a technical condition that is difficult to check. On the other hand, (C2) and (C3) have a certain normative appeal and are much easier to test. Indeed, (C2) is a one-dimensional assumption and means that if a lottery ν is at least as risky as μ shifted to the right by every arbitrarily small amount, then ν is also at least as risky as μ. To put (C3) into perspective, we note that it is considerably weaker than the following condition:

(C3′) If for $\mu \in \mathcal{M}_{1,c}$ there exists an $\eta \in \mathcal{M}_1$ such that for all $\lambda \in [0,1)$, one has $\lambda\mu + (1-\lambda)\eta \trianglerighteq \tau$ for some $\tau \in \mathcal{S}_\nu$, then $\mu \in S_\nu$,

which is a stronger version of the directional closedness assumption

(C3″) If $\mu, \eta \in \mathcal{M}_{1,c}$ are such that $\lambda\mu + (1-\lambda)\eta \in S_\nu$ for all $\lambda \in [0,1)$, then $\mu \in S_\nu$.

Remark 1.2.3 below shows that a subset $\mathcal{A}$ of a Banach lattice $(E, \ge)$ is norm-closed if it satisfies (C3″) and the monotonicity condition

(C4′) $\mu \ge \tau \in \mathcal{A}$ implies $\mu \in \mathcal{A}$.

However, $(\mathcal{M}_{1,c}, \trianglerighteq)$ is only a convex set with a partial order, and the topology $\sigma(\mathcal{M}_{1,c}, C)$ is not metrizable; see Remark 1.3.1. For our proof of Theorem 1.2.1 to work, conditions like (C3″) and (C4′) are not enough. It needs (C2) and (C3).

Theorem 1.2.1 *Every subset $\mathcal{A}$ of $\mathcal{M}_{1,c}$ satisfying* (C2) *and* (C3) *is $\sigma(\mathcal{M}_{1,c}, C)$-closed.*

The proof is given in Sect. 1.3. As a consequence, one obtains the following:

Theorem 1.2.2 *If all sublevel sets of $\succcurlyeq$ satisfy* (C1)–(C3), *then $\succcurlyeq$ has a numerical representation $\rho : \mathcal{M}_{1,c} \to [-\infty, \infty]$ satisfying* (A1)–(A3). *Moreover, for every such ρ, there exists a unique risk function R with properties* (R1)–(R5) *such that* (1.1) *holds.*

Proof By Theorem 1.2.1, the sublevel sets of $\succcurlyeq$ are closed in $\sigma(\mathcal{M}_{1,c}, C)$. Since they are also convex and monotone with respect to $\trianglerighteq$, the theorem follows from Drapeau and Kupper (2010, Theorem 3.5). □

Remark 1.2.3 A subset $\mathcal{A}$ of a Banach lattice $(E, \geq)$ satisfying (C3$''$) and (C4$'$) is norm-closed. Indeed, if x_n is a sequence in $\mathcal{A}$ converging to $x \in E$, one can pass to a subsequence and assume that $\|x_n - x\| \leq 2^{-n}/n$. For $y := x + \sum_{k=1}^{\infty} k(x_k - x)^+$ and $\lambda \in [0, 1)$, one then has

$$\lambda x + (1-\lambda)y = x + (1-\lambda)\sum_{k=1}^{\infty} k(x_k - x)^+ \geq x + (1-\lambda)n(x_n - x)^+ \geq x_n$$

for all $n \geq 1/(1-\lambda)$. Hence, $\lambda x + (1-\lambda)y \in \mathcal{A}$ for each $\lambda \in [0, 1)$, from which one obtains $x \in \mathcal{A}$.

Example 1.2.4 Value-at-Risk is a risk measure widely used in the banking industry. For a random variable X and a level $\alpha \in (0, 1)$, it is defined by

$$V@R_\alpha(X) = \inf\{x \in \mathbb{R} : P[X + x < 0] \leq \alpha\}$$

and gives the minimal amount of money which has to be added to X to keep the probability of default below α. It is well known that the sublevel sets of $V@R_\alpha$ are not convex; see, for instance, Artzner et al. (1999) or Föllmer and Schied (2004). However, it depends on X only through its distribution. So it can be defined on $\mathcal{M}_{1,c}$ by

$$V@R_\alpha(\mu) = -q_\mu^+(\alpha), \tag{1.2}$$

where q_μ^+ is the right-quantile function of μ given by

$$q_\mu^+ := \sup\{x \in \mathbb{R} : F_\mu(x) \leq \alpha\}.$$

As subsets of $\mathcal{M}_{1,c}$, the sublevel sets are convex. Moreover, it can easily be checked that they satisfy (C2) and (C3). So it follows from Theorem 1.2.2 that (1.2) has a robust representation of the form (1.1). Indeed, one has

$$V@R_\alpha(\mu) = \sup_{l \in L} -l^{-1}\left(\frac{\langle l, \mu\rangle - \alpha l(-\infty)}{1-\alpha}\right) = -\inf_{l \in L} l^{-1}\left(\frac{\langle l, \mu\rangle - \alpha l(-\infty)}{1-\alpha}\right),$$

where l^{-1} is the left-inverse of l; see Drapeau and Kupper (2010).

The two following examples show that none of conditions (C2) and (C3) can be dropped from the assumptions of Theorem 1.2.1.

Example 1.2.5 The set

$$\mathcal{A} := \{\mu \in \mathcal{M}_{1,c} : \mu_* > 0\}$$

is clearly not $\sigma(\mathcal{M}_{1,c}, C)$-closed since $\delta_{1/n} \in \mathcal{A}$ converges in $\sigma(\mathcal{M}_{1,c}, C)$ to $\delta_0 \notin \mathcal{A}$. However, it fulfills condition (C3). Indeed, if μ is an element of $\mathcal{M}_{1,c}$ such that for all $\lambda \in [0,1)$ and $\eta \in \mathcal{M}_1$ with $\eta_* \geq \mu^*$ and $\eta^* = +\infty$, one has $\lambda\mu + (1-\lambda)\eta \trianglerighteq \tau$ for some $\tau \in \mathcal{A}$, then $\mu_* > 0$, and therefore, $\mu \in \mathcal{A}$. By Theorem 1.2.1, $\mathcal{A}$ cannot fulfill condition (C2), which can also be seen directly by observing that $T_m\delta_0 \in \mathcal{A}$ for all $m > 0$ and $\delta_0 \notin \mathcal{A}$.

Example 1.2.6 Consider the set

$$\mathcal{A} := \left\{\mu \in \mathcal{M}_{1,c} : \mu^* \geq 2 \text{ and } \mu \trianglerighteq \left(1 - \frac{1}{n}\right)\delta_0 + \frac{1}{n}\delta_1 \text{ for some } n \geq 1\right\}.$$

It is not $\sigma(\mathcal{M}_{1,c}, C)$-closed since $(1-1/n)\delta_0 + 1/n\delta_2 \in \mathcal{A}$ converges in $\sigma(\mathcal{M}_{1,c}, C)$ to $\delta_0 \notin \mathcal{A}$. It can easily be seen that it fulfills (C2). Indeed, if $T_m\mu \in \mathcal{A}$ for all $m > 0$, then $\mu^* \geq 2$ and $\mu_* \geq 0$. Hence, $\mu \trianglerighteq (1-1/n)\delta_0 + 1/n\delta_1$ for some $n \geq 1$, and thus, $\mu \in \mathcal{A}$. It follows from Theorem 1.2.1 that (C3) cannot hold. In fact, δ_0 has the property that for all $\lambda \in [0,1)$ and $\eta \in \mathcal{M}_1$ satisfying $\eta_* \geq \delta_0^* = 0$ and $\eta^* = +\infty$, one can find a $\tau \in \mathcal{A}$ such that $\lambda\delta_0 + (1-\lambda)\eta \trianglerighteq \tau$. However, $\delta_0 \notin \mathcal{A}$ since $\delta_0^* = 0 < 2$.

1.3 Weak Closedness of Monotone Sets of Lotteries

Before giving the proof of Theorem 1.2.1, we compare the topology $\sigma(\mathcal{M}_{1,c}, C)$ to $\sigma(\mathcal{M}_{1,c}, C_b)$, where C_b denotes the space of all bounded continuous functions $f : \mathbb{R} \to \mathbb{R}$.

Remark 1.3.1 It is well known that the topology $\sigma(\mathcal{M}_1, C_b)$, and therefore also $\sigma(\mathcal{M}_{1,c}, C_b)$, is generated by the Lévy metric

$$d_L(\mu, \nu) := \inf\{\varepsilon > 0 : F_\mu(x - \varepsilon) - \varepsilon \leq F_\nu(x) \leq F_\mu(x + \varepsilon) + \varepsilon \text{ for all } x \in \mathbb{R}\}.$$

But $\sigma(\mathcal{M}_{1,c}, C)$ is finer than $\sigma(\mathcal{M}_{1,c}, C_b)$, which can easily be seen from the fact that $(1 - 1/n)\delta_0 + \delta_n/n$ converges to δ_0 in $\sigma(\mathcal{M}_{1,c}, C_b)$ but not in $\sigma(\mathcal{M}_{1,c}, C)$. Moreover, in contrast to $\sigma(\mathcal{M}_{1,c}, C_b)$, $\sigma(\mathcal{M}_{1,c}, C)$ is not metrizable. Indeed, if one assumes that $\sigma(\mathcal{M}_{1,c}, C)$ is generated by a metric, then for every ball $B_{1/n}(\nu)$ around a fixed $\nu \in \mathcal{M}_{1,c}$, there exist functions $u_1^n, \ldots, u_{i_n}^n$ in $C \setminus \{0\}$ such that

$$U_n := \{\mu : |\langle u_i^n, \mu - \nu\rangle| \leq 1, \ i = 1, \ldots, i_n\} \subset B_{1/n}(\nu).$$

By shifting, one can assume that $\nu^* = 0$. Define the function $u \in C$ by $u(x) = 0$ for $x \leq 0$. For $m = 1, 2, \dots$, set

$$u(m) = \max_{1\leq n\leq m} \max_{1\leq i\leq i_n} \left(\left|2u_i^n(m)\right| \vee m\right)$$

and interpolate linearly so that it becomes a continuous function $u : \mathbb{R} \to \mathbb{R}$. There must be an n such that

$$U_n \subset B_{1/n}(\nu) \subset \left\{\mu : \left|\langle u, \mu - \nu\rangle\right| \leq 1/2\right\}. \tag{1.3}$$

Choose $m \geq n$ such that

$$\frac{1}{u(m)}\left|\langle u_i^n, \nu\rangle\right| \leq 1/2 \quad \text{for all } i = 1, \dots, i_n.$$

Set $\lambda = 1/u(m)$ and $\mu = \lambda\delta_m + (1 - \lambda)\nu$. Then

$$\left|\langle u_i^n, \mu - \nu\rangle\right| = \lambda\left|\langle u_i^n, \delta_m - \nu\rangle\right| \leq \frac{|u_i^n(m)|}{u(m)} + \lambda\left|\langle u_i^n, \nu\rangle\right| \leq 1$$

for all $i = 1, \dots, i_n$. So μ is in U_n, but at the same time,

$$\langle u, \mu - \nu\rangle = \lambda\langle u, \delta_m - \nu\rangle = 1,$$

a contradiction to (1.3).

Proof of Theorem 1.2.1 Assume that (μ_α) is a net in $\mathcal{A}$ converging to some $\mu \in \mathcal{M}_{1,c}$ in $\sigma(\mathcal{M}_{1,c}, C)$. Fix $m > 0$, $\lambda \in [0, 1)$, and $\eta \in \mathcal{M}_1$ such that $T_m\mu^* \leq \eta_*$ and $\eta^* = +\infty$. Note that

$$\lambda F_\mu(x - m) = 0 \leq F_\alpha(x) \quad \text{for all } x < \mu_* + m \text{ and every } \alpha. \tag{1.4}$$

Set $b := (1 - \lambda) \wedge (m/2)$ and $c := F_\mu(\mu_* + b) > 0$. Since $\mu_\alpha \to \mu$ in $\sigma(\mathcal{M}_{1,c}, C_b)$, there exists α_0 such that

$$F_\mu(x - bc) - bc \leq F_\alpha(x) \quad \text{for all } x \in \mathbb{R} \text{ and } \alpha \geq \alpha_0.$$

For $x \geq \mu_* + m$, one has $F_\mu(x - bc) \geq c$, and therefore,

$$\lambda F_\mu(x - m) \leq \lambda F_\mu(x - bc) \leq F_\mu(x - bc) - bc \leq F_\alpha(x) \quad \text{for all } \alpha \geq \alpha_0. \tag{1.5}$$

It follows from (1.4)–(1.5) that

$$\lambda F_\mu(x - m) + (1 - \lambda)F_\eta(x) \leq F_\alpha(x) \quad \text{for all } \alpha \geq \alpha_0 \text{ and } x < \mu^* + m.$$

Now choose a nonnegative function $u \in C$ such that

$$u(x) = 0 \quad \text{for } x \leq \mu^* \quad \text{and} \quad u(x) \geq \frac{1}{(1 - \lambda)(1 - F_\eta(x))} \quad \text{for } x \geq \mu^* + m.$$

There exists an $\alpha \geq \alpha_0$ such that

$$\left|\langle u, \mu_\alpha - \mu\rangle\right| < 1,$$

which implies

$$\lambda F_\mu(x - m) + (1 - \lambda) F_\eta(x) \leq F_\alpha(x) \quad \text{for all } x \geq m + \mu^*.$$

Indeed, if there existed an $x_0 \geq \mu^* + m$ such that

$$\lambda F_\mu(x_0 - m) + (1 - \lambda) F_\eta(x_0) > F_\alpha(x_0),$$

it would follow that

$$\langle u, \mu_\alpha - \mu\rangle = \int u \, d\mu_\alpha \geq u(x_0)(1 - \lambda)\big(1 - F_\eta(x_0)\big) \geq 1,$$

a contradiction. So we have shown that

$$\lambda T_m \mu + (1 - \lambda)\eta \trianglerighteq \mu_\alpha.$$

It follows from (C3) that $T_m \mu \in \mathcal{A}$ for all $m > 0$, which by (C2), implies $\mu \in \mathcal{A}$. □

Acknowledgements P. Cheridito was supported in part by NSF Grant DMS-0642361. S. Drapeau financial support from MATHEON project E.11 is gratefully acknowledged.

Chapter 2
Multivariate Concave and Convex Stochastic Dominance

Michel Denuit, Louis Eeckhoudt, Ilia Tsetlin, and Robert L. Winkler

Keywords Decision analysis: multiple criteria, risk · Group decisions · Utility/preference: multiattribute utility, stochastic dominance, stochastic orders

2.1 Introduction

One of the big challenges in decision analysis is the assessment of a decision maker's utility function. To the extent that the alternatives under consideration in a decision-making problem can be partially ordered based on less-than-full information about the utility function, the problem can be simplified somewhat by eliminating dominated alternatives. At the same time, partial orders can help in the creation of alternatives by providing an indication of the types of strategies that might be most promising. Stochastic dominance has been studied extensively in the univariate case, particularly in the finance and economics literature; early papers are Hadar and Russell (1969) and Hanoch and Levy (1969). For example, assuming that util-

M. Denuit
Institut des Sciences Actuarielles & Institut de Statistique, Université Catholique de Louvain, Rue des Wallons 6, 1348 Louvain-la-Neuve, Belgium
e-mail: michel.denuit@uclouvain.be

L. Eeckhoudt
IESEG School of Management, LEM, Université Catholique de Lille, Lille, France

L. Eeckhoudt
CORE, Université Catholique de Louvain, Voie du Roman Pays 34, 1348 Louvain-la-Neuve, Belgium
e-mail: eeckhoudt@fucam.ac.be

I. Tsetlin (✉)
INSEAD, 1 Ayer Rajah Avenue, Singapore 138676, Singapore
e-mail: ilia.tsetlin@insead.edu

R.L. Winkler
Fuqua School of Business, Duke University, Durham, NC 27708-0120, USA
e-mail: rwinkler@duke.edu

F. Biagini et al. (eds.), *Risk Measures and Attitudes*, EAA Series,
DOI 10.1007/978-1-4471-4926-2_2, © Springer-Verlag London 2013

ity for money is increasing and concave can simplify many problems in finance and economics.

Moreover, stochastic dominance can be even more helpful in group decision making, where the challenge is amplified by divergent preferences. Even though the group members can be expected to have different utility functions, these utility functions may share some common characteristics. Thus, if an alternative can be eliminated based on an individual's utility function being risk averse, then all group members will agree that it can be eliminated if each member of the group is risk averse, even though the degree of risk aversion may vary considerably within the group.

Multiattribute consequences make the assessment of utility even more difficult, and extensions to multivariate stochastic dominance are tricky because there are many multivariate stochastic orders (Denuit et al. 1999; Müller and Stoyan 2002; Shaked and Shantikumar 2007; Denuit and Mesfioui 2010) on which the dominance can be based. Hazen (1986) investigates multivariate stochastic dominance when simple forms of utility independence (Keeney and Raiffa 1976) can be assumed. If utility independence cannot be assumed, the potential benefits of stochastic dominance are even greater. Studies of multivariate stochastic dominance include Levy and Paroush (1974), Levhari et al. (1975), Mosler (1984), Scarsini (1988), and Denuit and Eeckhoudt (2010). In this paper we use a stochastic order that can be related to characteristics such as risk aversion and correlation aversion, is consistent with a basic preference assumption, and is a natural extension of the standard order typically used for univariate stochastic dominance. We also consider a stochastic order that is consistent with characteristics such as risk taking and correlation loving by reversing the basic preference assumption.

The objective of this paper is to study multivariate stochastic dominance for the above-mentioned stochastic orders. In Sect. 2.2, we define these stochastic orders, which form the basis for what we call nth-degree multivariate concave and convex stochastic dominance. We extend the concept of nth-degree risk to the multivariate case and show that it is related to multivariate concave and convex stochastic dominance. Then we show a connection with a preference for combining good with bad in the concave case and with the opposite preference for combining good with good and bad with bad in the convex case. We develop some ways to facilitate the comparison of alternatives via multivariate stochastic dominance in Sect. 2.3, focusing on the impact of background risk and on eliminating alternatives from consideration by comparing an alternative with a mixture of other alternatives. A simple hypothetical example is presented to illustrate the concepts from Sects. 2.2–2.3. In Sect. 2.4, we consider infinite-degree concave and convex stochastic dominance, which can be related to utility functions that are mixtures of multiattribute exponential utilities, and present dominance results when the joint probability distribution for the attributes is multivariate normal. In Sect. 2.5, we compare our multivariate stochastic dominance with dominance based on another family of stochastic orders possessing some interesting similarities and differences. A brief summary and concluding comments are given in Sect. 2.6.

2.2 Multivariate Stochastic Dominance

2.2.1 Multivariate Concave and Convex Stochastic Dominance

We begin by defining some notation. A random vector is denoted by a tilde, $\tilde{\mathbf{x}}$, and $\mathbf{0}$ is a vector of zeroes. For two N-dimensional vectors $\mathbf{x}$ and $\mathbf{y}$, $\mathbf{x} > \mathbf{y}$ if $x_j > y_j$ for $j = 1, \ldots, N$ and $\mathbf{x} \gneqq \mathbf{y}$ if $x_j \geq y_j$ for all j and $\mathbf{x} \neq \mathbf{y}$. Also, $\mathbf{x} + \mathbf{y}$ denotes the component-wise sum, $(x_1 + y_1, \ldots, x_N + y_N)$.

Next, we consider a differentiable utility function u for a vector of N attributes and formalize the notion of alternating signs for the partial derivatives of u.

Definition 2.2.1

$$\underline{\mathbb{U}}_n^N = \left\{ u \middle| (-1)^{k-1} \frac{\partial^k u(\mathbf{x})}{\partial x_{i_1} \cdots \partial x_{i_k}} \geq 0 \text{ for } k = 1, \ldots, n \text{ and } i_j \in \{1, \ldots, N\}, j = 1, \ldots, k \right\}.$$

$\underline{\mathbb{U}}_n^N$ consists of all N-dimensional real-valued functions for which all partial derivatives of a given degree up to degree n have the same sign, and that sign alternates, being positive for odd degrees and negative for even degrees. Observe that if $u \in \underline{\mathbb{U}}_n^N$, then $u \in \underline{\mathbb{U}}_k^N$ for any $k < n$. Also, if $u \in \underline{\mathbb{U}}_n^N$, then for any $k < n$ and $i_j \in \{1, \ldots, N\}$, $j = 1, \ldots, k$,

$$(-1)^k \frac{\partial^k u(\mathbf{x})}{\partial x_{i_1} \partial x_{i_2} \cdots \partial x_{i_k}} \in \underline{\mathbb{U}}_{n-k}^N .$$

Now we use $\underline{\mathbb{U}}_n^N$ to define multivariate concave stochastic dominance.

Definition 2.2.2 For random vectors $\tilde{\mathbf{x}}$ and $\tilde{\mathbf{y}}$ with support contained in $[\underline{\mathbf{x}}, \bar{\mathbf{x}}]$, $\tilde{\mathbf{x}}$ dominates $\tilde{\mathbf{y}}$ in the sense of nth-degree concave stochastic dominance if

$$\mathbb{E}\big[u(\tilde{\mathbf{x}})\big] \geq \mathbb{E}\big[u(\tilde{\mathbf{y}})\big]$$

for all $u \in \underline{\mathbb{U}}_n^N$, u defined on $[\underline{\mathbf{x}}, \bar{\mathbf{x}}]$.

Next we define multivariate convex stochastic dominance.

Definition 2.2.3

$$\overline{\mathbb{U}}_n^N = \left\{ u \middle| \frac{\partial^k u(\mathbf{x})}{\partial x_{i_1} \cdots \partial x_{i_k}} \geq 0 \text{ for } k = 1, \ldots, n \text{ and } i_j \in \{1, \ldots, N\}, j = 1, \ldots, k \right\}.$$

$\overline{\mathbb{U}}_n^N$, consisting of all N-dimensional real-valued functions for which all partial derivatives of degree up to n are positive, is called $\mathbb{U}_{s\text{-idircx}}$ by Denuit and Mesfioui (2010) and forms the basis for the s-increasing directionally convex order. Similar to $\underline{\mathbb{U}}_n^N$, if $u \in \overline{\mathbb{U}}_n^N$, then $u \in \overline{\mathbb{U}}_k^N$ for any $k < n$. Also, if $u \in \overline{\mathbb{U}}_n^N$, then for any $k < n$ and $i_j \in \{1, \ldots, N\}$, $j = 1, \ldots, k$,

$$\frac{\partial^k u(\mathbf{x})}{\partial x_{i_1} \partial x_{i_2} \cdots \partial x_{i_k}} \in \overline{\mathbb{U}}_{n-k}^N .$$

Definition 2.2.4 For random vectors $\tilde{\mathbf{x}}$ and $\tilde{\mathbf{y}}$ with support contained in $[\underline{\mathbf{x}}, \bar{\mathbf{x}}]$, $\tilde{\mathbf{x}}$ dominates $\tilde{\mathbf{y}}$ in the sense of nth-degree convex stochastic dominance if

$$\mathbb{E}\big[u(\tilde{\mathbf{x}})\big] \geq \mathbb{E}\big[u(\tilde{\mathbf{y}})\big]$$

for all $u \in \overline{\mathbb{U}}_n^N$, u defined on $[\underline{\mathbf{x}}, \bar{\mathbf{x}}]$.

Remark 2.2.5 The multivariate convex stochastic dominance in Definition 2.2.4 is different from what Fishburn (1974) calls convex stochastic dominance. Fishburn's usage of "convex" does not relate to the utility function. Instead, it refers to dominance results for convex combinations, or mixtures, of probability distributions in the univariate case, which we will extend to the multivariate case in Sect. 2.3.2 and use to eliminate alternatives in decision-making problems in Sect. 2.3.3. To clarify the distinction, we will use the term "mixture dominance" when referring to the type of stochastic dominance developed by Fishburn (1974, 1978). In contrast, our multivariate convex stochastic dominance can be thought of as "risk-seeking stochastic dominance" because $u \in \overline{\mathbb{U}}_n^N$ for any $n > 1$ implies that u is risk seeking with respect to each individual attribute and is multivariate risk seeking in the sense of Richard (1975). Similarly, our multivariate concave stochastic dominance from Definition 2.2.2 can be thought of as "risk-averse stochastic dominance" because $u \in \underline{\mathbb{U}}_n^N$ for any $n > 1$ means that u is risk averse with respect to each attribute and is multivariate risk averse (Richard 1975). The correlation-increasing transformations of Epstein and Tanny (1980) link multivariate risk aversion and multivariate risk seeking to correlation aversion and correlation loving, respectively.

2.2.2 The Notion of nth-Degree Risk in the Multivariate Case

By Definition 2.2.2 (2.2.4), concave (convex) stochastic dominance of degree n implies stochastic dominance of any higher degree. To isolate a higher-degree effect in the univariate case, Ekern (1980) introduced the concept of nth-degree risk. Examples include Rothschild and Stiglitz (1970), who focus on a 2nd-degree effect in terms of a mean-preserving spread, and Menezes et al. (1980), who isolate a 3rd-degree effect via a mean-variance-preserving transformation. This subsection extends that concept to the multivariate case and relates it to concave and convex stochastic dominance.

Definition 2.2.6 For random vectors $\tilde{\mathbf{x}}$ and $\tilde{\mathbf{y}}$ with support contained in $[\underline{\mathbf{x}}, \bar{\mathbf{x}}]$, $-\infty < \underline{\mathbf{x}} < \bar{\mathbf{x}} < \infty$, $\tilde{\mathbf{y}}$ has more nth-degree risk than $\tilde{\mathbf{x}}$ if

$$\mathbb{E}\big[u(\tilde{\mathbf{x}})\big] \geq \mathbb{E}\big[u(\tilde{\mathbf{y}})\big]$$

for all u defined on $[\underline{\mathbf{x}}, \bar{\mathbf{x}}]$ such that

$$(-1)^{n-1}\frac{\partial^n u(\mathbf{x})}{\partial x_{i_1}\cdots\partial x_{i_n}} \geq 0$$

for any $i_j \in \{1, \ldots, N\}$, $j = 1, \ldots, n$.

Theorem 2.2.7 *The random vector $\tilde{\mathbf{y}}$ has more nth-degree risk than the random vector $\tilde{\mathbf{x}}$ if and only if*

(1) *$\tilde{\mathbf{x}}$ dominates $\tilde{\mathbf{y}}$ in the sense of nth-degree concave stochastic dominance, and*
(2) *the kth moments of $\tilde{\mathbf{x}}$ and $\tilde{\mathbf{y}}$ are identical for $k = 1, \ldots, n-1$:*

$$\mathbb{E}[\tilde{x}_{i_1}\tilde{x}_{i_2}\cdots\tilde{x}_{i_k}] = \mathbb{E}[\tilde{y}_{i_1}\tilde{y}_{i_2}\cdots\tilde{y}_{i_k}]$$

for any $i_j \in \{1, \ldots, N\}$, $j = 1, \ldots, k$.

Proof For the "only if" part, (1) holds by the definition of $\underline{\mathbb{U}}_n^N$. For (2), consider $u(\mathbf{x}) = x_{i_1}x_{i_2}\cdots x_{i_k}$ for any $i_j \in \{1, \ldots, N\}$ and $k < n$. For this $u(\mathbf{x})$,

$$(-1)^{n-1}\frac{\partial^n u(\mathbf{x})}{\partial x_{i_1}\cdots\partial x_{i_n}} = 0$$

for any $i_j \in \{1, \ldots, N\}$, $j = 1, \ldots, n$. Therefore,

$$\mathbb{E}[\tilde{x}_{i_1}\tilde{x}_{i_2}\cdots\tilde{x}_{i_k}] \geq \mathbb{E}[\tilde{y}_{i_1}\tilde{y}_{i_2}\cdots\tilde{y}_{i_k}].$$

Similarly, for $u(\mathbf{x}) = -x_{i_1}x_{i_2}\cdots x_{i_k}$,

$$\mathbb{E}[\tilde{y}_{i_1}\tilde{y}_{i_2}\cdots\tilde{y}_{i_k}] \geq \mathbb{E}[\tilde{x}_{i_1}\tilde{x}_{i_2}\cdots\tilde{x}_{i_k}].$$

Thus,

$$\mathbb{E}[\tilde{x}_{i_1}\tilde{x}_{i_2}\cdots\tilde{x}_{i_k}] = \mathbb{E}[\tilde{y}_{i_1}\tilde{y}_{i_2}\cdots\tilde{y}_{i_k}].$$

Now, suppose that (1) and (2) hold. We need to prove that for any u such that

$$(-1)^{n-1}\frac{\partial^n u(\mathbf{x})}{\partial x_{i_1}\cdots\partial x_{i_n}} \geq 0$$

for any $i_j \in \{1, \ldots, N\}$, $j = 1, \ldots, n$, $\mathbb{E}[u(\tilde{\mathbf{x}})] \geq \mathbb{E}[u(\tilde{\mathbf{y}})]$. Since u is differentiable at least n times, all lower-degree derivatives exist and are bounded on $[\underline{\mathbf{x}}, \bar{\mathbf{x}}]$. Therefore, there exist coefficients $c_{i_1,\dots,i_k}$ for $k = 1, \dots, n-1$ and any $i_j \in \{1, \dots, N\}$, $j = 1, \dots, k$, such that

$$v(\mathbf{x}) = u(\mathbf{x}) + \sum c_{i_1,\dots,i_k} x_{i_1} x_{i_2} \cdots x_{i_k}$$

and $v \in \underline{\mathbb{U}}_n^N$, where the summation is over all possible combinations of $i_1, \dots, i_k$. By (1), $\mathbb{E}[v(\tilde{\mathbf{x}})] \geq \mathbb{E}[v(\tilde{\mathbf{y}})]$, and by (2), $\mathbb{E}[v(\tilde{\mathbf{x}})] - \mathbb{E}[v(\tilde{\mathbf{y}})] = \mathbb{E}[u(\tilde{\mathbf{x}})] - \mathbb{E}[u(\tilde{\mathbf{y}})]$. Therefore, $\mathbb{E}[u(\tilde{\mathbf{x}})] \geq \mathbb{E}[u(\tilde{\mathbf{y}})]$. □

Remark 2.2.8 In the univariate case, Ekern (1980) defines a person as being "nth-degree risk averse" if the nth derivative of her utility function is positive (negative) when n is odd (even). Our interpretation of multivariate concave stochastic dominance as risk-averse stochastic dominance is consistent with the extension of the notion of being nth-degree risk averse to the multivariate case.

Theorem 2.2.9 *The random vector $\tilde{\mathbf{y}}$ has more nth-degree risk than the random vector $\tilde{\mathbf{x}}$ if and only if*

(1) *$\tilde{\mathbf{x}}$ dominates $\tilde{\mathbf{y}}$ ($\tilde{\mathbf{y}}$ dominates $\tilde{\mathbf{x}}$) in the sense of nth-degree convex stochastic dominance when n is odd (even), and*
(2) *the kth moments of $\tilde{\mathbf{x}}$ and $\tilde{\mathbf{y}}$ are identical for $k = 1, \dots, n-1$.*

The proof of Theorem 2.2.9 is similar to the proof of Theorem 2.2.7.

Corollary 2.2.10 (to Theorems 2.2.7 and 2.2.9) *If n is odd (even) and the kth moments of $\tilde{\mathbf{x}}$ and $\tilde{\mathbf{y}}$ are identical for $k = 1, \dots, n-1$, then $\tilde{\mathbf{x}}$ dominates $\tilde{\mathbf{y}}$ in the sense of nth-degree concave stochastic dominance if and only if $\tilde{\mathbf{x}}$ dominates $\tilde{\mathbf{y}}$ ($\tilde{\mathbf{y}}$ dominates $\tilde{\mathbf{x}}$) in the sense of nth-degree convex stochastic dominance.*

Thus, if all moments of degree less than n are identical, convex dominance goes along with higher nth moments for both odd and even n. With concave dominance, this holds only for odd n. For even n, concave dominance goes along with lower nth moments. These results relate stochastic dominance to ordering by moments, in the sense that convex dominance likes all moments to be higher, whereas concave dominance likes odd moments to be higher and even moments to be lower.

2.2.3 Connections with Preferences for Combining Good with Bad or Good with Good and Bad with Bad

Next, we show a connection between our definition of multivariate concave stochastic dominance and a preference for combining good lotteries with bad lotteries as opposed to combining good lotteries with good and bad lotteries with bad. This preference can be thought of as a type of risk aversion, so it is similar in spirit to the assumption of risk aversion in the single-attribute case. We let $\langle \tilde{\mathbf{x}}, \tilde{\mathbf{y}} \rangle$ denote a lottery with equal chances of getting $\tilde{\mathbf{x}}$ or $\tilde{\mathbf{y}}$.

Theorem 2.2.11 *Let $\tilde{\mathbf{x}}_m$, $\tilde{\mathbf{y}}_m$, $\tilde{\mathbf{x}}_n$, and $\tilde{\mathbf{y}}_n$ be mutually independent N-dimensional random vectors with $\tilde{\mathbf{x}}_i$ dominating $\tilde{\mathbf{y}}_i$ in the sense of ith-degree concave stochastic dominance, $i = m, n$. Then $\langle \tilde{\mathbf{x}}_m + \tilde{\mathbf{y}}_n, \tilde{\mathbf{y}}_m + \tilde{\mathbf{x}}_n \rangle$ dominates $\langle \tilde{\mathbf{x}}_m + \tilde{\mathbf{x}}_n, \tilde{\mathbf{y}}_m + \tilde{\mathbf{y}}_n \rangle$ in the sense of $(n+m)$th-degree concave stochastic dominance.*

Proof Consider any $u \in \underline{\mathbb{U}}^N_{n+m}$, and denote

$$v(\mathbf{z}) = \mathbb{E}\big[u(\tilde{\mathbf{y}}_m + \mathbf{z})\big] - \mathbb{E}\big[u(\tilde{\mathbf{x}}_m + \mathbf{z})\big].$$

Now

$$0.5\,\mathbb{E}\big[u(\tilde{\mathbf{x}}_m + \tilde{\mathbf{y}}_n)\big] + 0.5\,\mathbb{E}\big[u(\tilde{\mathbf{y}}_m + \tilde{\mathbf{x}}_n)\big] \geq 0.5\,\mathbb{E}\big[u(\tilde{\mathbf{x}}_m + \tilde{\mathbf{x}}_n)\big] + 0.5\,\mathbb{E}\big[u(\tilde{\mathbf{y}}_m + \tilde{\mathbf{y}}_n)\big]$$

is equivalent to

$$\mathbb{E}\big[u(\tilde{\mathbf{y}}_m + \tilde{\mathbf{x}}_n)\big] - \mathbb{E}\big[u(\tilde{\mathbf{x}}_m + \tilde{\mathbf{x}}_n)\big] \geq \mathbb{E}\big[u(\tilde{\mathbf{y}}_m + \tilde{\mathbf{y}}_n)\big] - \mathbb{E}\big[u(\tilde{\mathbf{x}}_m + \tilde{\mathbf{y}}_n)\big],$$

or $\mathbb{E}[v(\tilde{\mathbf{x}}_n)] \geq \mathbb{E}[v(\tilde{\mathbf{y}}_n)]$. It remains to show that $v(\mathbf{z}) \in \underline{\mathbb{U}}^N_n$. For any $k = 1, \ldots, n$ and any $i_j \in \{1, \ldots, N\}$, $j = 1, \ldots, k$,

$$(-1)^{k-1}\frac{\partial^k v(\mathbf{z})}{\partial z_{i_1}\cdots\partial z_{i_k}} = (-1)^{k-1}\left(\mathbb{E}\left[\frac{\partial^k u(\tilde{\mathbf{y}}_m + \mathbf{z})}{\partial z_{i_1}\cdots\partial z_{i_k}}\right] - \mathbb{E}\left[\frac{\partial^k u(\tilde{\mathbf{x}}_m + \mathbf{z})}{\partial z_{i_1}\cdots\partial z_{i_k}}\right]\right),$$

and

$$(-1)^k\frac{\partial^k u(\mathbf{x})}{\partial x_{i_1}\partial x_{i_2}\cdots\partial x_{i_k}} \in \underline{\mathbb{U}}^N_{m+n-k} \subset \underline{\mathbb{U}}^N_m.$$

Therefore, $(-1)^{k-1}\frac{\partial^k v(\mathbf{z})}{\partial z_{i_1}\cdots\partial z_{i_k}} \geq 0$, so $v(\mathbf{z}) \in \underline{\mathbb{U}}^N_n$. □

Theorem 2.2.11 shows that concave stochastic dominance from Definition 2.2.2 is consistent with a preference for combining good with bad (up to degree n), where good and bad are understood in terms of lower-degree concave stochastic dominance. What if a decision maker prefers to combine good with good and bad with bad, as opposed to combining good with bad?

Theorem 2.2.12 *Let $\tilde{\mathbf{x}}_m$, $\tilde{\mathbf{y}}_m$, $\tilde{\mathbf{x}}_n$, and $\tilde{\mathbf{y}}_n$ be mutually independent N-dimensional random vectors with $\tilde{\mathbf{x}}_i$ dominating $\tilde{\mathbf{y}}_i$ in the sense of ith-degree convex stochastic dominance, $i = m, n$. Then $\langle \tilde{\mathbf{x}}_m + \tilde{\mathbf{x}}_n, \tilde{\mathbf{y}}_m + \tilde{\mathbf{y}}_n \rangle$ dominates $\langle \tilde{\mathbf{x}}_m + \tilde{\mathbf{y}}_n, \tilde{\mathbf{y}}_m + \tilde{\mathbf{x}}_n \rangle$ in the sense of $(n+m)$th-degree convex stochastic dominance.*

Proof This is, essentially, a corollary to Theorem 2.2.11. Observe that $u(\mathbf{x}) \in \underline{\mathbb{U}}^N_n$ if and only if $-u(\underline{\mathbf{x}} + \bar{\mathbf{x}} - \mathbf{x}) \in \overline{\mathbb{U}}^N_n$. Therefore, $\tilde{\mathbf{x}}_i$ dominates $\tilde{\mathbf{y}}_i$ in the sense of ith-degree convex stochastic dominance if and only if $\underline{\mathbf{x}} + \bar{\mathbf{x}} - \tilde{\mathbf{y}}_i$ dominates $\underline{\mathbf{x}} + \bar{\mathbf{x}} - \tilde{\mathbf{x}}_i$ in the sense of ith-degree concave stochastic dominance. By Theorem 2.2.11, $\langle \underline{\mathbf{x}} + \bar{\mathbf{x}} - \tilde{\mathbf{x}}_m + \underline{\mathbf{x}} + \bar{\mathbf{x}} - \tilde{\mathbf{y}}_n, \underline{\mathbf{x}} + \bar{\mathbf{x}} - \tilde{\mathbf{y}}_m + \underline{\mathbf{x}} + \bar{\mathbf{x}} - \tilde{\mathbf{x}}_n \rangle$ dominates $\langle \underline{\mathbf{x}} + \bar{\mathbf{x}} - \tilde{\mathbf{x}}_m + \underline{\mathbf{x}} + \bar{\mathbf{x}} -$

$\tilde{\mathbf{x}}_n, \underline{\mathbf{x}} + \bar{\mathbf{x}} - \tilde{\mathbf{y}}_m + \underline{\mathbf{x}} + \bar{\mathbf{x}} - \tilde{\mathbf{y}}_n\rangle$ in the sense of $(n+m)$th-degree concave stochastic dominance, and thus $\langle \tilde{\mathbf{x}}_m + \tilde{\mathbf{x}}_n, \tilde{\mathbf{y}}_m + \tilde{\mathbf{y}}_n\rangle$ dominates $\langle \tilde{\mathbf{x}}_m + \tilde{\mathbf{y}}_n, \tilde{\mathbf{y}}_m + \tilde{\mathbf{x}}_n\rangle$ in the sense of $(n+m)$th-degree convex stochastic dominance. □

Definition 2.2.2 extends the standard definition of univariate stochastic dominance to the multivariate case. As Theorem 2.2.11 shows, it preserves a preference for combining good with bad (Eeckhoudt and Schlesinger 2006; Eeckhoudt et al. 2009). The preference for combining good with bad associated with $u \in \underline{\mathbb{U}}_n^N$ can be viewed as a form of risk aversion. For example, it implies that u is correlation averse (Epstein and Tanny 1980; Eeckhoudt et al. 2007, Denuit et al. 2010), which can be interpreted as a form of risk aversion. Definition 2.2.4 and Theorem 2.2.12 develop similar orderings based on the opposite preference for combining good with good and bad with bad, and show the connection between convex and concave stochastic dominance that follows from the fact that $u(\mathbf{x}) \in \underline{\mathbb{U}}_n^N$ if and only if $-u(\underline{\mathbf{x}} + \bar{\mathbf{x}} - \mathbf{x}) \in \overline{\mathbb{U}}_n^N$. The preference for combining good with good and bad with bad associated with $u \in \overline{\mathbb{U}}_n^N$ implies that u is correlation loving, a form of risk taking.

2.3 Comparing Alternatives via Multivariate Stochastic Dominance

Here we present several results that are useful for comparing alternatives according to the stochastic dominance relations from Sect. 2.2. In Sect. 2.3.1 we show conditions under which dominance orderings remain unchanged in the presence of background risk, with independence playing an important role. In Sect. 2.3.2 we use mixture dominance to show that an alternative, even if not dominated by any single alternative, can be eliminated from consideration if it is dominated by a mixture of other alternatives. A simple example is presented in Sect. 2.3.3 to illustrate the concepts from Sects. 2.2–2.3.

2.3.1 Stochastic Dominance with Additive and Multiplicative Background Risk

When one faces a choice between two (or more) risky alternatives, this decision is often not made in isolation, in the sense that there are other risks that affect the decision maker but are outside of the decision maker's control. Therefore, it is important to know whether a stochastic dominance ordering established in the absence of background risk will remain the same when background risk is present.

Consider a choice between two projects, with consequences characterized by random vectors $\tilde{\mathbf{x}}$ and $\tilde{\mathbf{y}}$. In the presence of additive background risk, represented by the random vector $\tilde{\mathbf{a}}$, we are interested in comparing $\tilde{\mathbf{a}} + \tilde{\mathbf{x}}$ and $\tilde{\mathbf{a}} + \tilde{\mathbf{y}}$. In the presence of multiplicative background risk, represented by the random vector $\tilde{\mathbf{m}}$, the appropriate comparison is between $\tilde{\mathbf{m}} \otimes \tilde{\mathbf{x}}$ and $\tilde{\mathbf{m}} \otimes \tilde{\mathbf{y}}$, where $\mathbf{m} \otimes \mathbf{x}$ denotes the component-wise product, $(m_1 x_1, \ldots, m_N x_N)$. If both additive and multiplicative background risks are present, $\tilde{\mathbf{a}} + \tilde{\mathbf{m}} \otimes \tilde{\mathbf{x}}$ and $\tilde{\mathbf{a}} + \tilde{\mathbf{m}} \otimes \tilde{\mathbf{y}}$ are compared.

Theorem 2.3.1 *Let* $\tilde{\mathbf{x}}$, $\tilde{\mathbf{y}}$, $\tilde{\mathbf{a}}$, *and* $\tilde{\mathbf{m}}$, $\tilde{\mathbf{m}} \gneqq \mathbf{0}$, *be* N*-dimensional random vectors such that for any fixed* $\mathbf{a}$ *and* $\mathbf{m}$, $\tilde{\mathbf{x}}|\mathbf{m}, \mathbf{a}$ *dominates* $\tilde{\mathbf{y}}|\mathbf{m}, \mathbf{a}$ *in the sense of* n*th-degree concave (convex) stochastic dominance. Then* $\tilde{\mathbf{a}} + \tilde{\mathbf{m}} \otimes \tilde{\mathbf{x}}$ *dominates* $\tilde{\mathbf{a}} + \tilde{\mathbf{m}} \otimes \tilde{\mathbf{y}}$ *in the sense of* n*th-degree concave (convex) stochastic dominance.*

Proof Consider any $u \in \underline{\mathbb{U}}_n^N$ $(u \in \overline{\mathbb{U}}_n^N)$. For any fixed $\mathbf{a}$ and $\mathbf{m}$, $v(\mathbf{x} \mid \mathbf{a}, \mathbf{m}) = u(\mathbf{a} + \mathbf{m} \otimes \mathbf{x})$, as a function of $\mathbf{x}$, belongs to $\underline{\mathbb{U}}_n^N$ $(\overline{\mathbb{U}}_n^N)$. Therefore, $\mathbb{E}[v(\tilde{\mathbf{x}} \mid \mathbf{a}, \mathbf{m})] \geq \mathbb{E}[v(\tilde{\mathbf{y}} \mid \mathbf{a}, \mathbf{m})]$. Taking expectations with respect to $\tilde{\mathbf{a}}$ and $\tilde{\mathbf{m}}$ yields $\mathbb{E}[u(\tilde{\mathbf{a}} + \tilde{\mathbf{m}} \otimes \tilde{\mathbf{x}})] \geq \mathbb{E}[u(\tilde{\mathbf{a}} + \tilde{\mathbf{m}} \otimes \tilde{\mathbf{y}})]$. □

The result of Theorem 2.3.1 is quite intuitive. If $\tilde{\mathbf{x}}$ is preferred to $\tilde{\mathbf{y}}$ for each possible value of $\mathbf{a}$ and $\mathbf{m}$, then $\tilde{\mathbf{x}}$ is preferred to $\tilde{\mathbf{y}}$ even if we are uncertain about the exact values of $\mathbf{a}$ and $\mathbf{m}$. If the project risk is independent of the background risk, the situation is further simplified.

Corollary 2.3.2 (to Theorem 2.3.1) *Let* $\tilde{\mathbf{x}}$, $\tilde{\mathbf{y}}$, $\tilde{\mathbf{a}}$, *and* $\tilde{\mathbf{m}}$, $\tilde{\mathbf{m}} \gneqq \mathbf{0}$, *be* N*-dimensional random vectors such that* $\tilde{\mathbf{x}}$ *and* $\tilde{\mathbf{y}}$ *are independent of* $\tilde{\mathbf{a}}$ *and* $\tilde{\mathbf{m}}$. *If* $\tilde{\mathbf{x}}$ *dominates* $\tilde{\mathbf{y}}$ *in the sense of* n*th-degree concave (convex) stochastic dominance, then* $\tilde{\mathbf{a}} + \tilde{\mathbf{m}} \otimes \tilde{\mathbf{x}}$ *dominates* $\tilde{\mathbf{a}} + \tilde{\mathbf{m}} \otimes \tilde{\mathbf{y}}$ *in the sense of* n*th-degree concave (convex) stochastic dominance.*

Thus, independent background risk preserves stochastic dominance orderings. Note that no assumption is made about the relationship between the background risks $\tilde{\mathbf{a}}$ and $\tilde{\mathbf{m}}$; they can be dependent. The assumption of independence of the project risk and the background risk is crucial, however. If background risk is not independent of project risk, preferences with and without background risk might be the opposite (Tsetlin and Winkler 2005). For example, suppose that a manager is considering adding a new project to an existing portfolio of projects. Let $\tilde{\mathbf{x}}$ and $\tilde{\mathbf{y}}$ represent the consequences of two potential new projects, and let $\tilde{\mathbf{a}}$ represent the consequences of the existing portfolio. Even if the manager is multivariate risk averse and $\tilde{\mathbf{x}}$ dominates $\tilde{\mathbf{y}}$ in terms of multivariate concave stochastic dominance, she might prefer the new project associated with $\tilde{\mathbf{y}}$ (i.e., prefer $\tilde{\mathbf{a}} + \tilde{\mathbf{y}}$ to $\tilde{\mathbf{a}} + \tilde{\mathbf{x}}$) if the correlations between the components of $\tilde{\mathbf{a}}$ and $\tilde{\mathbf{y}}$ are smaller than those for $\tilde{\mathbf{a}}$ and $\tilde{\mathbf{x}}$.

Theorem 2.3.1 and its Corollary 2.3.2 can also be used to compare random vectors that are functions of other random vectors, which can be ordered by stochastic dominance. For instance, if the consequences of a particular alternative can be represented as $\tilde{\mathbf{a}} + \tilde{\mathbf{m}} \otimes \tilde{\mathbf{x}}$ and any of the mutually independent random vectors $\tilde{\mathbf{x}}$, $\tilde{\mathbf{a}}$, and $\tilde{\mathbf{m}}$ is improved in the sense of stochastic dominance, what can we say about the resulting changes to this alternative?

Corollary 2.3.3 (to Theorem 2.3.1) *Let* $\tilde{\mathbf{x}}_1, \tilde{\mathbf{y}}_1, \tilde{\mathbf{x}}_2$, *and* $\tilde{\mathbf{y}}_2$ *be mutually independent* N*-dimensional random vectors with* $\tilde{\mathbf{x}}_i$ *dominating* $\tilde{\mathbf{y}}_i$ *in the sense of* n*th-degree concave (convex) stochastic dominance,* $i = 1, 2$. *Then* $\tilde{\mathbf{x}}_1 + \tilde{\mathbf{x}}_2$ *dominates* $\tilde{\mathbf{y}}_1 + \tilde{\mathbf{y}}_2$ *in the sense of* n*th-degree concave (convex) stochastic dominance. If* $\tilde{\mathbf{x}}_1 \gneqq \mathbf{0}$, $\tilde{\mathbf{y}}_1 \gneqq \mathbf{0}$, $\tilde{\mathbf{x}}_2 \gneqq \mathbf{0}$, *and* $\tilde{\mathbf{y}}_2 \gneqq \mathbf{0}$, *then* $\tilde{\mathbf{x}}_1 \otimes \tilde{\mathbf{x}}_2$ *dominates* $\tilde{\mathbf{y}}_1 \otimes \tilde{\mathbf{y}}_2$ *in the sense of* n*th-degree concave (convex) stochastic dominance.*

Remark 2.3.4 It might be that, e.g., $\tilde{\mathbf{x}}_1 + \tilde{\mathbf{x}}_2$ dominates $\tilde{\mathbf{y}}_1 + \tilde{\mathbf{y}}_2$ in the sense of stochastic dominance of degree lower than n. For example, consider the univariate case (i.e., $N = 1$) with $\tilde{x}_1 = 1$, $\tilde{x}_2 = \tilde{y}_1 = 0$, and $\tilde{y}_2 = [-c, c]$. Then $\tilde{x}_i$ dominates $\tilde{y}_i$ in the sense of second-degree concave stochastic dominance for $i = 1, 2$, but also note that $\tilde{x}_1$ dominates $\tilde{y}_1$ in the sense of first-degree stochastic dominance. In this case $\tilde{x}_1 + \tilde{x}_2 = 1$ and $\tilde{y}_1 + \tilde{y}_2 = [-c, c]$. For $c \leq 1$, $\tilde{x}_1 + \tilde{x}_2$ dominates $\tilde{y}_1 + \tilde{y}_2$ in the sense of first-degree stochastic dominance, but for $c > 1$, $\tilde{x}_1 + \tilde{x}_2$ dominates $\tilde{y}_1 + \tilde{y}_2$ only in the sense of second-degree concave stochastic dominance.

Theorem 2.3.1 and its corollaries show that, e.g., adding a nonnegative random vector improves a multivariate distribution in the sense of first-degree concave and convex stochastic dominance. They also imply that if a set of N variables can be divided into two stochastically independent subgroups and one of these groups is improved in the sense of nth-degree concave (convex) stochastic dominance, then the joint distribution over all N variables is improved in the sense of nth-degree concave (convex) stochastic dominance. In particular, if N random variables are independent, then their joint distribution is improved in the sense of nth-degree concave (convex) stochastic dominance whenever the marginal distribution of any of the variables is improved in the sense of nth-degree concave (convex) stochastic dominance.

2.3.2 Elimination by Mixtures

If an alternative (represented by a random vector) is dominated by some other alternative when the decision maker's utility falls in a particular class (e.g., $u \in \underline{\mathbb{U}}_n^N$ for concave stochastic dominance and $u \in \overline{\mathbb{U}}_n^N$ for convex stochastic dominance), then the dominated alternative can be eliminated from further consideration, thereby simplifying the decision-making problem. Mixture dominance, developed by Fishburn (1974) as "convex stochastic dominance" for the univariate case, allows us to eliminate an alternative even if it is not dominated by any other single alternative, as long as it is dominated by a mixture of other alternatives, which is a weaker condition (Fishburn 1978). We define mixture dominance for the multivariate case and then extend Fishburn's (1978) result regarding elimination by mixtures.

Definition 2.3.5 For the random vectors $\tilde{\mathbf{x}}_1, \ldots, \tilde{\mathbf{x}}_k$ and utility class $\mathbb{U}^*$, $\tilde{\mathbf{x}}_{-k} = (\tilde{\mathbf{x}}_1, \ldots, \tilde{\mathbf{x}}_{k-1})$ dominates $\tilde{\mathbf{x}}_k$ in the sense of mixture dominance with respect to $\mathbb{U}^*$

if there exists $\mathbf{p} = (p_1, \ldots, p_{k-1}) \geq 0$, $\sum_{i=1}^{k-1} p_i = 1$, such that

$$\sum_{i=1}^{k-1} p_i \, \mathbb{E}\big[u(\tilde{\mathbf{x}}_i)\big] \geq \mathbb{E}\big[u(\tilde{\mathbf{x}}_k)\big]$$

for all $u \in \mathbb{U}^*$.

From Definition 2.3.5, the mixture can be thought of as a two-step process. In the first step, an alternative (a random vector $\tilde{\mathbf{x}}_i$) is chosen from $\tilde{\mathbf{x}}_{-k}$ where p_i represents the probability of choosing $\tilde{\mathbf{x}}_i$. Then at the second step, the uncertainty about $\tilde{\mathbf{x}}_i$ is resolved. Mixture dominance means that this mixture has a higher expected utility than $\tilde{\mathbf{x}}_k$ for all $u \in \mathbb{U}^*$.

Theorem 2.3.6 *If $\tilde{\mathbf{x}}_{-k}$ dominates $\tilde{\mathbf{x}}_k$ in the sense of mixture dominance with respect to $\mathbb{U}^*$, then for every $u \in \mathbb{U}^*$, there is an $i \in \{1, \ldots, k-1\}$ such that*

$$\mathbb{E}\big[u(\tilde{\mathbf{x}}_i)\big] \geq \mathbb{E}\big[u(\tilde{\mathbf{x}}_k)\big].$$

Proof For any $u \in \mathbb{U}^*$, there is a $\mathbf{p}$ such that

$$\sum_{i=1}^{k-1} p_i \, \mathbb{E}\big[u(\tilde{\mathbf{x}}_i)\big] \geq \mathbb{E}\big[u(\tilde{\mathbf{x}}_k)\big].$$

This is impossible unless $\mathbb{E}[u(\tilde{\mathbf{x}}_i)] \geq \mathbb{E}[u(\tilde{\mathbf{x}}_k)]$ for some $i \in \{1, \ldots, k-1\}$. □

Note that the $\tilde{\mathbf{x}}_i$ in Theorem 2.3.6 can be different for different $u \in \mathbb{U}^*$. The importance of Theorem 2.3.6 is that if $u \in \mathbb{U}^*$ and $\tilde{\mathbf{x}}_{-k}$ dominates $\tilde{\mathbf{x}}_k$ in the sense of mixture dominance with respect to the utility class of interest, then we can eliminate $\tilde{\mathbf{x}}_k$ from consideration even if none of $\tilde{\mathbf{x}}_1, \ldots, \tilde{\mathbf{x}}_{k-1}$ dominates $\tilde{\mathbf{x}}_k$ individually. Reducing the set of alternatives that need to be considered seriously is always helpful. Since some of the mixing probabilities can be zero, we can eliminate an alternative if it is dominated in the sense of mixture dominance by any subset of the other alternatives. Of course, mixture dominance with respect to $\underline{\mathbb{U}}_n^N$ or $\overline{\mathbb{U}}_n^N$ is of particular interest because it invokes concave or convex stochastic dominance and relates to a preference for combining good with bad or the opposite preference for combining good with good and bad with bad.

2.3.3 Example

A decision-making task is somewhat simplified if some potential alternatives can be eliminated from consideration without having to assess the full utility function, and that is where multivariate stochastic dominance can be helpful. In this section, we present a simple hypothetical example to illustrate the concepts from Sects. 2.2–2.3 without getting distracted by complicating details.

Suppose that a telecom company is entering a new market and deciding among different entry strategies. For simplicity, assume that a decision maker (DM) focuses on two attributes, x_1 (the net present value (NPV) of profits for the first five years, in millions of dollars) and x_2 (the market share in percentage terms at the end of the five-year period). To begin, it is not surprising to find that the DM prefers more of each of these attributes to less. For example, she prefers $(x_1, x_2) = (300, 40)$ to $(200, 30)$. This is simple first-degree multivariate stochastic dominance.

Next, if the DM concludes that she is risk averse with respect to NPV, then $(250, 30)$ would be preferred to $\langle(300, 30), (200, 30)\rangle$, a risky alternative that yields $(300, 30)$ or $(200, 30)$ with equal probabilities. Similarly, if she is risk averse with respect to market share, then $(250, 35)$ would be preferred to $\langle(250, 30), (250, 40)\rangle$. These two choices are consistent with second-degree concave stochastic dominance but not sufficient to indicate that she would always want to behave in accordance with second-degree concave stochastic dominance. For example, the risk aversion with respect to NPV and market share is not sufficient to dictate her choice between the two risky alternatives $\langle(300, 40), (200, 30)\rangle$ and $\langle(300, 30), (200, 40)\rangle$. She states a preference for the latter and decides after some thought that she is, in general, correlation averse. Thus, her preferences are consistent with second-degree concave stochastic dominance.

In practice, most comparisons between competing alternatives are not as clear-cut as the above examples. In other words, once obviously inferior alternatives have been eliminated, it may be hard to find many cases where one alternative dominates another. However, by looking at three or more alternatives, we may still be able to eliminate alternatives via mixture dominance, as discussed in Sect. 2.3.2.

For a simple example, consider the choice among three alternatives: $(300, 30)$, $(200, 40)$, and $\langle(300, 40), (200, 30)\rangle$. The first alternative gives a higher NPV, the second alternative gives a higher market share, and the third alternative is risky, with equal chances of either the high NPV and the high market share or the low NPV and the low market share. Note that a 50–50 mixture of the first two alternatives, $\langle(300, 30), (200, 40)\rangle$ dominates the third alternative by second-degree concave stochastic dominance, consistent with the DM's preference for combining good with bad. By Theorem 2.3.6, then, we can eliminate the third alternative.

Of course, if the DM has the opposite preference for combining good with good and bad with bad, then convex stochastic dominance is relevant, and the second-degree dominance orderings in the above examples will be reversed. For example, $\langle(300, 30), (200, 30)\rangle$ dominates $(250, 30)$ by second-degree convex stochastic dominance. Similarly, $\langle(300, 40), (200, 30)\rangle$ dominates $\langle(300, 30), (200, 40)\rangle$ by second-degree convex stochastic dominance, reflecting the fact that the DM is correlation loving.

The above comparisons among alternatives might have to be made in the presence of background risk. For example, the DM might be uncertain about the financial results of other ongoing projects of the telecom company, implying additive background risk with respect to the first attribute (NPV). She might also be uncertain about competitors' moves, which could translate into additive background risk with respect to the second attribute (market share). Finally, suppose that the company operates internationally and wants to express its NPV in another currency. In this case, the appropriate exchange rate, in the absence of hedging, would operate as multiplicative background risk with respect to the first attribute. As shown in Sect. 2.3.1, if the consequences of each alternative are independent of the background risk, then any stochastic dominance orderings are preserved, and any resulting elimination of alternatives remains optimal under such background risk.

2.4 Infinite-Degree Dominance

Now we explore what emerges if a preference between combining good with bad, or combining good with good and bad with bad, holds for any n. In this case dominance relations are defined via $\underline{\mathbb{U}}_\infty^N$ and $\overline{\mathbb{U}}_\infty^N$ that extend $\underline{\mathbb{U}}_n^N$ and $\overline{\mathbb{U}}_n^N$.

Definition 2.4.1

$$\underline{\mathbb{U}}_\infty^N = \left\{ u \middle| (-1)^{k-1} \frac{\partial^k u(\mathbf{x})}{\partial x_{i_1} \cdots \partial x_{i_k}} \geq 0 \text{ for } k = 1, 2, \ldots \text{ and } i_j \in \{1, \ldots, N\}, j = 1, \ldots, k \right\},$$

and

$$\overline{\mathbb{U}}_\infty^N = \left\{ u \middle| \frac{\partial^k u(\mathbf{x})}{\partial x_{i_1} \cdots \partial x_{i_k}} \geq 0 \text{ for } k = 1, 2, \ldots \text{ and } i_j \in \{1, \ldots, N\}, j = 1, \ldots, k \right\}.$$

Definition 2.4.2 For random vectors $\tilde{\mathbf{x}}$ and $\tilde{\mathbf{y}}$ with support contained in $[\underline{\mathbf{x}}, \bar{\mathbf{x}}]$, $\tilde{\mathbf{x}}$ dominates $\tilde{\mathbf{y}}$ in the sense of infinite-degree concave (convex) stochastic dominance if

$$\mathbb{E}\big[u(\tilde{\mathbf{x}})\big] \geq \mathbb{E}\big[u(\tilde{\mathbf{y}})\big]$$

for all $u \in \underline{\mathbb{U}}_\infty^N$ ($u \in \overline{\mathbb{U}}_\infty^N$), u defined on $[\underline{\mathbf{x}}, \bar{\mathbf{x}}]$.

Increasing the degree of dominance (n) restricts the set of utility functions with respect to which two random vectors are compared. Similarly, expanding the domain of definition of u (i.e., decreasing $\underline{\mathbf{x}}$ and/or increasing $\bar{\mathbf{x}}$) also restricts the set of utility functions and thus increases the set of random vectors that can be ordered by stochastic dominance.

2.4.1 Infinite-Degree Dominance and Mixtures of Multiattribute Exponential Utilities

We show in Theorem 2.4.3 that any $u \in \underline{\mathbb{U}}_\infty^N$, u defined on $[\underline{\mathbf{x}}, \infty)$, or $u \in \overline{\mathbb{U}}_\infty^N$, u defined on $(-\infty, \bar{\mathbf{x}}]$, is a mixture of multiattribute exponential utilities. Theorem 2.4.4

then shows that infinite-order dominance can be operationalized via multiattribute exponential utilities.

Theorem 2.4.3 *Consider a function $u(\mathbf{x})$ defined on $[\underline{\mathbf{x}}, \infty)$. Then $u \in \underline{\mathbb{U}}_{\infty}^{N}$ if and only if there exists a (not necessarily finite) measure F on $[\mathbf{0}, \infty)$ and constants $b_1, \ldots, b_N$ with $b_i \geq 0$, $i = 1, \ldots, N$, such that*

$$\begin{aligned} u(\mathbf{x}) &= u(\underline{\mathbf{x}}) \\ &\quad + \int_0^{\infty} \cdots \int_0^{\infty} \big(1 - \exp\big(-\big(r_1(x_1 - \underline{x}_1) + \cdots \\ &\quad + r_N(x_N - \underline{x}_N)\big)\big)\big)\, dF(r_1, \ldots, r_N) + \sum_{i=1}^{N} b_i(x_i - \underline{x}_i). \end{aligned} \tag{2.1}$$

Viewing the linear terms in (2.1) *as limiting forms of exponential utilities (as $r_i \to 0$ with $r_j = 0$ for $j \neq i$) and rescaling, we can express any $u \in \underline{\mathbb{U}}_{\infty}^{N}$, u defined on $[\underline{\mathbf{x}}, \infty)$, as a mixture of multiattribute exponential utilities,*

$$u(\mathbf{x}) = -\int_0^{\infty} \cdots \int_0^{\infty} \exp(-r_1 x_1 - \cdots - r_N x_N)\, dF(r_1, \ldots, r_N). \tag{2.2}$$

Similarly, any $u \in \overline{\mathbb{U}}_{\infty}^{N}$, u defined on $(-\infty, \bar{\mathbf{x}}]$, can be expressed as

$$u(\mathbf{x}) = \int_0^{\infty} \cdots \int_0^{\infty} \exp(r_1 x_1 + \cdots + r_N x_N)\, dF(r_1, \ldots, r_N). \tag{2.3}$$

A proof for the concave case in Theorem 2.4.3 is given in Tsetlin and Winkler (2009), and the proof for the convex case is similar. From Theorem 2.4.3 we can state the following result without a proof.

Theorem 2.4.4 *The random vector $\tilde{\mathbf{x}}$ dominates the random vector $\tilde{\mathbf{y}}$ in the sense of infinite-degree concave stochastic dominance for u defined on $[\underline{\mathbf{x}}, \infty)$ if and only if*

$$\mathbb{E}\big[\exp(-r_1\tilde{y}_1 - \cdots - r_N\tilde{y}_N)\big] \geq \mathbb{E}\big[\exp(-r_1\tilde{x}_1 - \cdots - r_N\tilde{x}_N)\big]$$

for all $\mathbf{r} \in [\mathbf{0}, \infty)$, and $\tilde{\mathbf{x}}$ dominates $\tilde{\mathbf{y}}$ in the sense of infinite-degree convex stochastic dominance for u defined on $(-\infty, \bar{\mathbf{x}}]$ if and only if

$$\mathbb{E}\big[\exp(r_1\tilde{x}_1 + \cdots + r_N\tilde{x}_N)\big] \geq \mathbb{E}\big[\exp(r_1\tilde{y}_1 + \cdots + r_N\tilde{y}_N)\big]$$

for all $\mathbf{r} \in [\mathbf{0}, \infty)$.

Theorem 2.4.4 provides a convenient criterion for comparing multivariate probability distributions. Note that the expectations in Theorem 2.4.4 correspond to moment generating functions for distributions of $\tilde{\mathbf{x}}$ and $\tilde{\mathbf{y}}$. If we define $M_{\tilde{\mathbf{x}}}(\mathbf{r}) = \mathbb{E}[\exp(r_1\tilde{x}_1 + \cdots + r_N\tilde{x}_N)]$, then for concave stochastic dominance, we need

$M_{\tilde{\mathbf{x}}}(\mathbf{r}) \leq M_{\tilde{\mathbf{y}}}(\mathbf{r})$ for all $\mathbf{r} \in (-\infty, \mathbf{0}]$, and for convex stochastic dominance, we need $M_{\tilde{\mathbf{x}}}(\mathbf{r}) \geq M_{\tilde{\mathbf{y}}}(\mathbf{r})$ for all $\mathbf{r} \in [\mathbf{0}, \infty)$.

Remark 2.4.5 The domain of definition of u is crucial for the result stated in Theorem 2.4.4. For instance, if $\tilde{\mathbf{x}} = (x_1, x_2) = (0.5, 0.5)$ and $\tilde{\mathbf{y}} = \langle(0, 1), (1, 0)\rangle$, then by examining the expectations in Theorem 2.4.4 we can show that $\tilde{\mathbf{x}}$ dominates $\tilde{\mathbf{y}}$ by infinite-degree concave stochastic dominance for u defined on $[\underline{\mathbf{x}}, \infty)$ (e.g., on $[\mathbf{0}, \infty)$). However, consider $u(\mathbf{x}) = x_1 + x_2 - x_1 x_2$, u defined on $[\mathbf{0}, \mathbf{1}]$. Theorem 2.4.4 does not apply here, and taking expectations with respect to u yields $\mathbb{E}[u(\tilde{\mathbf{x}})] = 0.75 < \mathbb{E}[u(\tilde{\mathbf{y}})] = 1$. Therefore, $\tilde{\mathbf{x}}$ does not dominate $\tilde{\mathbf{y}}$ by infinite-degree concave stochastic dominance. If we increase the upper limit of the domain of this u above $\mathbf{1}$, then $u \notin \underline{\mathbb{U}}^2_\infty$ because $\frac{\partial u(\mathbf{x})}{\partial x_i} < 0$, $i = 1, 2$, when $\mathbf{x} > \mathbf{1}$. A similar situation can occur for any N, including the univariate case ($N = 1$). As noted previously, expanding the domain of definition of u restricts the set of utility functions with respect to which random vectors are compared. In the example, the set of utility functions $u \in \underline{\mathbb{U}}^2_\infty$ defined on $[\mathbf{0}, \mathbf{1}]$ is larger than the set of utility functions $u \in \underline{\mathbb{U}}^2_\infty$ defined on $[\mathbf{0}, \infty)$. The former set includes $u(\mathbf{x}) = x_1 + x_2 - x_1 x_2$, whereas the latter does not.

2.4.2 Comparison of Multivariate Normal Distributions via Infinite-Degree Dominance

The multivariate normal distribution is the most commonly encountered multivariate distribution, is very tractable, and is a reasonable representation of uncertainty in many situations. Müller (2001) provides several results on the stochastic ordering of multivariate normal distributions. The expectations appearing in Theorem 2.4.4 are especially tractable in this case, and thus the comparison of two multivariate normal distributions based on infinite-degree (concave and convex) stochastic dominance is greatly simplified. If the random vector $\tilde{\mathbf{x}}$ is multivariate normal with mean vector $\boldsymbol{\mu} = (\mu_1, \ldots, \mu_N)$ and covariance matrix $\boldsymbol{\Sigma} = (\rho_{ij}\sigma_i\sigma_j)$, then

$$\mathbb{E}\big[\exp(r_1\tilde{x}_1 + \cdots + r_N\tilde{x}_N)\big] = \exp\left(\mathbf{r}\boldsymbol{\mu}^t + \left(\frac{\mathbf{r}\boldsymbol{\Sigma}\mathbf{r}^t}{2}\right)\right),$$

where a superscript t denotes transposition, and

$$\mathbf{r}\boldsymbol{\mu}^t + \left(\frac{\mathbf{r}\boldsymbol{\Sigma}\mathbf{r}^t}{2}\right) = \sum_{i=1}^{N} r_i\mu_i + \left(\sum_{i=1}^{N}\sum_{j=1}^{N} r_i r_j \frac{\rho_{ij}\sigma_i\sigma_j}{2}\right).$$

Thus, we have the following corollary to Theorem 2.4.4.

Corollary 2.4.6 (to Theorem 2.4.4) *Let $\tilde{\mathbf{x}}$ and $\tilde{\mathbf{y}}$ be multivariate normal vectors with mean vectors $\boldsymbol{\mu}_x$ and $\boldsymbol{\mu}_y$, and covariance matrices $\boldsymbol{\Sigma}_x$ and $\boldsymbol{\Sigma}_y$. Then $\tilde{\mathbf{x}}$ dominates*

$\tilde{\mathbf{y}}$ *in the sense of infinite-degree concave stochastic dominance if and only if*

$$-\mathbf{r}\boldsymbol{\mu}_y^t + \left(\frac{\mathbf{r}\boldsymbol{\Sigma}_y\mathbf{r}^t}{2}\right) \geq -\mathbf{r}\boldsymbol{\mu}_x^t + \left(\frac{\mathbf{r}\boldsymbol{\Sigma}_x\mathbf{r}^t}{2}\right)$$

for all $\mathbf{r} \in [\mathbf{0}, \infty)$, *and* $\tilde{\mathbf{x}}$ *dominates* $\tilde{\mathbf{y}}$ *in the sense of infinite-degree convex stochastic dominance if and only if*

$$\mathbf{r}\boldsymbol{\mu}_x^t + \left(\frac{\mathbf{r}\boldsymbol{\Sigma}_x\mathbf{r}^t}{2}\right) \geq \mathbf{r}\boldsymbol{\mu}_y^t + \left(\frac{\mathbf{r}\boldsymbol{\Sigma}_y\mathbf{r}^t}{2}\right)$$

for all $\mathbf{r} \in [\mathbf{0}, \infty)$.

Thus, increasing any mean μ_i leads to stochastic dominance improvement (both concave and convex). Decreasing any correlation ρ_{ij} leads to concave (convex) stochastic dominance improvement (deterioration). Decreasing any standard deviation σ_i leads to concave (convex) stochastic dominance improvement (deterioration) if $\rho_{ij} \geq 0$ for all j. However, if $\rho_{ij} < 0$ for some j, things are more complicated. Overall, adding independent noise to attribute i leads to the increase of σ_i and to the decrease of the absolute value of correlations ρ_{ij}. Thus, increasing σ_i without changing correlations is equivalent to adding independent noise to attribute i and then to adjusting the correlations ρ_{ij} up (if ρ_{ij} is positive) or down (if ρ_{ij} is negative). For concave (convex) stochastic dominance, adding independent noise is bad (good), and adjusting correlations up (down) is bad (good). If all correlations are positive, increasing any standard deviation leads to convex (concave) stochastic dominance improvement (deterioration). If some correlations are negative, the effect might go either way. Tsetlin and Winkler (2007) established similar confounding effects of increasing standard deviations in target-oriented situations.

2.5 Comparisons with Other Multivariate Stochastic Orders

Many multivariate stochastic orders have been studied, and the appropriate order upon which to base multivariate stochastic dominance is not as obvious as it is in the univariate case. Once we move from $N = 1$ to $N > 1$, the relationship among the attributes complicates matters both in terms of the joint probability distribution and in terms of the utility function.

Two commonly used multivariate stochastic orders are the lower and upper orthant orders, based on lower orthants $\{\mathbf{x} \mid \mathbf{x} \leq \mathbf{c}\}$ and upper orthants $\{\mathbf{x} \mid \mathbf{x} > \mathbf{c}\}$ for a given $\mathbf{c}$ (Müller and Stoyan 2002). By definition, $\tilde{\mathbf{x}}$ dominates $\tilde{\mathbf{y}}$ via the lower orthant order if

$$\mathbb{P}(\tilde{\mathbf{x}} \leq \mathbf{c}) \leq \mathbb{P}(\tilde{\mathbf{y}} \leq \mathbf{c})$$

for all $\mathbf{c} \in [\underline{\mathbf{x}}, \bar{\mathbf{x}}]$, and $\tilde{\mathbf{x}}$ dominates $\tilde{\mathbf{y}}$ via the upper orthant order if

$$\mathbb{P}(\tilde{\mathbf{x}} > \mathbf{c}) \geq \mathbb{P}(\tilde{\mathbf{y}} > \mathbf{c})$$

for all $\mathbf{c} \in [\underline{\mathbf{x}}, \bar{\mathbf{x}}]$ These orders highlight an important way in which moving from the univariate to the multivariate case makes stochastic orders and stochastic dominance more complex. In the univariate case, $\mathbb{P}(\tilde{x} \le c) + \mathbb{P}(\tilde{x} > c) = 1$ for any c. When $N \ge 2$, $\mathbb{P}(\tilde{\mathbf{x}} \le \mathbf{c}) + \mathbb{P}(\tilde{\mathbf{x}} > \mathbf{c}) \le 1$ for any $\mathbf{c} \in [\underline{\mathbf{x}}, \bar{\mathbf{x}}]$, and this becomes more of an issue as N increases because the lower and upper orthants for a given $\mathbf{c}$ represent only 2 of the 2^N orthants associated with $\mathbf{c}$.

We focus here on multivariate **s**-increasing orders, a family of stochastic orders for which some interesting connections and comparisons with our multivariate concave and convex stochastic dominance can be drawn. This helps to highlight potential advantages and disadvantages of our approach.

We begin by presenting the multivariate **s**-increasing concave order, where $\mathbf{s} = (s_1, \ldots, s_N)$ is a vector of positive integers, and defining stochastic dominance in terms of this order. This is a natural generalization of the bivariate (s_1, s_2)-increasing concave orders introduced by Denuit et al. (1999) and studied by Denuit and Eeckhoudt (2010) and Denuit et al. (2010).

Definition 2.5.1

$$\mathbb{U}^{\mathrm{N}}_{\mathbf{s}\text{-icv}} = \left\{ u \,\middle|\, (-1)^{\sum_{i=1}^{N} k_i - 1} \frac{\partial^{\sum_{i=1}^{N} k_i} u(\mathbf{x})}{\partial x_1^{k_1} \cdots \partial x_N^{k_N}} \ge 0 \text{ for } k_i = 0, 1, \ldots, s_i, \right.$$
$$\left. i = 1, \ldots, N, \sum_{i=1}^{N} k_i \ge 1 \right\}.$$

Definition 2.5.2 For random vectors $\tilde{\mathbf{x}}$ and $\tilde{\mathbf{y}}$ with support contained in $[\underline{\mathbf{x}}, \bar{\mathbf{x}}]$, $\tilde{\mathbf{x}}$ dominates $\tilde{\mathbf{y}}$ in the sense of the multivariate **s**-increasing concave order if

$$\mathbb{E}\big[u(\tilde{\mathbf{x}})\big] \ge \mathbb{E}\big[u(\tilde{\mathbf{y}})\big]$$

for all $u \in \mathbb{U}^{\mathrm{N}}_{\mathbf{s}\text{-icv}}$, u defined on $[\underline{\mathbf{x}}, \bar{\mathbf{x}}]$.

If $s_1 = \cdots = s_N = s$, we say that the order is an s-increasing concave order. Special cases of this are the lower orthant order when $s = 1$ and the lower orthant concave order when $s = 2$ (Mosler 1984).

Our multivariate concave stochastic dominance, based on $\underline{\mathbb{U}}_n^N$, has a convex counterpart, based on $\overline{\mathbb{U}}_n^N$. Similarly, $\mathbb{U}^{\mathrm{N}}_{\mathbf{s}\text{-icv}}$ and dominance in terms of the **s**-increasing concave order have convex counterparts (Denuit and Mesfioui 2010).

Definition 2.5.3

$$\mathbb{U}^{\mathrm{N}}_{\mathbf{s}\text{-icx}} = \left\{ u \,\middle|\, \frac{\partial^{\sum_{i=1}^{N} k_i} u(\mathbf{x})}{\partial x_1^{k_1} \cdots \partial x_N^{k_N}} \ge 0 \text{ for } k_i = 0, 1, \ldots, s_i, i = 1, \ldots, N, \sum_{i=1}^{N} k_i \ge 1 \right\}.$$

Definition 2.5.4 For random vectors $\tilde{\mathbf{x}}$ and $\tilde{\mathbf{y}}$ with support contained in $[\underline{\mathbf{x}}, \bar{\mathbf{x}}]$, $\tilde{\mathbf{x}}$ dominates $\tilde{\mathbf{y}}$ in the sense of the multivariate **s**-increasing convex order if

$$\mathbb{E}\big[u(\tilde{\mathbf{x}})\big] \geq \mathbb{E}\big[u(\tilde{\mathbf{y}})\big]$$

for all $u \in \mathbb{U}^{\mathbf{N}}_{\mathbf{s}\text{-icx}}$, u defined on $[\underline{\mathbf{x}}, \bar{\mathbf{x}}]$.

The **s**-increasing concave order and the **s**-increasing convex order are closely related, because $\tilde{\mathbf{x}}$ dominates $\tilde{\mathbf{y}}$ in the **s**-increasing concave order if and only if $\underline{\mathbf{x}} + \bar{\mathbf{x}} - \tilde{\mathbf{y}}$ dominates $\underline{\mathbf{x}} + \bar{\mathbf{x}} - \tilde{\mathbf{x}}$ in the **s**-increasing convex order. This follows from the fact that if $u \in \mathbb{U}^{\mathbf{N}}_{\mathbf{s}\text{-icv}}$, then $-u(\underline{\mathbf{x}} + \bar{\mathbf{x}} - \mathbf{x}) \in \mathbb{U}^{\mathbf{N}}_{\mathbf{s}\text{-icx}}$. An **s**-increasing convex order with $s_1 = \cdots = s_N = s$ is an s-increasing convex order. Analogous to the concave case, the s-increasing convex order with $s = 1$ is the upper orthant order.

Theorem 2.5.5 provides conditions characterizing stochastic dominance in the sense of the multivariate **s**-increasing concave and convex orders via partial moments, without reference to utilities. The following remark indicates an alternative characterization in terms of integral conditions.

Theorem 2.5.5 *Let* $\tilde{\mathbf{x}}$ *and* $\tilde{\mathbf{y}}$ *be random vectors with support contained in* $[\underline{\mathbf{x}}, \bar{\mathbf{x}}]$, $-\infty < \underline{\mathbf{x}} < \bar{\mathbf{x}} < \infty$, *and denote* $x_+ = \max\{x, 0\}$. *Then*

(1) $\tilde{\mathbf{x}}$ *dominates* $\tilde{\mathbf{y}}$ *in the sense of the multivariate* **s**-*increasing concave order if and only if*

$$\mathbb{E}\left[\prod_{i=1}^{N}(c_i - \tilde{x}_i)_+^{k_i-1}\right] \leq \mathbb{E}\left[\prod_{i=1}^{N}(c_i - \tilde{y}_i)_+^{k_i-1}\right]$$

for all $c_i \in [\underline{x}_i, \bar{x}_i]$ *if* $k_i = s_i$ *and* $c_i = \bar{x}_i$ *if* $k_i = 1, \ldots, s_i - 1$, $i = 1, \ldots, N$.

(2) $\tilde{\mathbf{x}}$ *dominates* $\tilde{\mathbf{y}}$ *in the sense of the multivariate* **s**-*increasing convex order if and only if*

$$\mathbb{E}\left[\prod_{i=1}^{N}(\tilde{x}_i - c_i)_+^{k_i-1}\right] \geq \mathbb{E}\left[\prod_{i=1}^{N}(\tilde{y}_i - c_i)_+^{k_i-1}\right]$$

for all $c_i \in [\underline{x}_i, \bar{x}_i]$ *if* $k_i = s_i$ *and* $c_i = \underline{x}_i$ *if* $k_i = 1, \ldots, s_i - 1$, $i = 1, \ldots, N$.

Proof Statement (2) is proven in Denuit and Mesfioui (2010) (Proposition 3.1). Statement (1) follows from (2) and the duality between the concave and convex orders: $\tilde{\mathbf{x}}$ dominates $\tilde{\mathbf{y}}$ in the sense of the multivariate **s**-increasing concave order if and only if $\underline{\mathbf{x}} + \bar{\mathbf{x}} - \tilde{\mathbf{y}}$ dominates $\underline{\mathbf{x}} + \bar{\mathbf{x}} - \tilde{\mathbf{x}}$ in the multivariate **s**-increasing convex order. Therefore, from (2),

$$\mathbb{E}\left[\prod_{i=1}^{N}(\underline{x}_i + \bar{x}_i - \tilde{y}_i - c_i)_+^{k_i-1}\right] \geq \mathbb{E}\left[\prod_{i=1}^{N}(\underline{x}_i + \bar{x}_i - \tilde{x}_i - c_i)_+^{k_i-1}\right]$$

with $c_i = \underline{x}_i$ if $k_i < s_i$ and $c_i \in [\underline{x}_i, \bar{x}_i]$ if $k_i = s_i$, which is equivalent to (1). □

Remark 2.5.6 Alternative necessary and sufficient conditions for dominance in the multivariate **s**-increasing concave and convex orders involve integral conditions. Let $F_{\tilde{\mathbf{x}}}$ be the cumulative distribution function $\mathbb{P}(\tilde{\mathbf{x}} \le \mathbf{x})$ for $\tilde{\mathbf{x}}$. Starting with $F_{\tilde{\mathbf{x}}}^{(1,\dots,1)} = F_{\tilde{\mathbf{x}}}$, define recursively the integrated left tails of $\tilde{\mathbf{x}}$ as

$$F_{\tilde{\mathbf{x}}}^{(k_1,\dots,k_i+1,\dots,k_N)}(\mathbf{x}) = \int_{\underline{x}_i}^{x_i} F_{\tilde{\mathbf{x}}}^{(k_1,\dots,k_i,\dots,k_N)}(x_1,\dots,z_i,\dots,x_N)\,dz_i \tag{2.4}$$

for $k_1,\dots,k_N \ge 1$. The lower partial moments in Theorem 2.5.5(1) can be expressed via integrated left tails:

$$\mathbb{E}\left[\prod_{i=1}^{N}(c_i - \tilde{x}_i)_+^{k_i-1}\right] = \left[\prod_{i=1}^{N}(k_i-1)!\right] F_{\tilde{\mathbf{x}}}^{(k_1,\dots,k_N)}(\mathbf{c}).$$

Then $\tilde{\mathbf{x}}$ dominates $\tilde{\mathbf{y}}$ in the sense of the multivariate **s**-increasing concave order if and only if $F_{\tilde{\mathbf{x}}}^{(k_1,\dots,k_N)}(\mathbf{c}) \le F_{\tilde{\mathbf{y}}}^{(k_1,\dots,k_N)}(\mathbf{c})$ for all $c_i \in [\underline{x}_i, \bar{x}_i]$ if $k_i = s_i$ and $c_i = \bar{x}_i$ if $k_i = 1,\dots,s_i-1$, $i = 1,\dots,N$. When $N = 1$, (2.4) is the standard integral condition for univariate stochastic dominance.

An expression similar to (2.4), involving integrated right tails of $\tilde{\mathbf{x}}$, holds for the multivariate **s**-increasing convex order (Denuit and Mesfioui 2010). If $G_{\tilde{\mathbf{x}}}(\mathbf{x}) = \mathbb{P}(\tilde{\mathbf{x}} > \mathbf{x})$ and $G_{\tilde{\mathbf{x}}}^{(1,\dots,1)} = G_{\tilde{\mathbf{x}}}$, define recursively

$$G_{\tilde{\mathbf{x}}}^{(k_1,\dots,k_i+1,\dots,k_N)}(\mathbf{x}) = \int_{x_i}^{\bar{x}_i} G_{\tilde{\mathbf{x}}}^{(k_1,\dots,k_i,\dots,k_N)}(x_1,\dots,z_i,\dots,x_N)\,dz_i \tag{2.5}$$

for $k_1,\dots,k_N \ge 1$. Then $\tilde{\mathbf{x}}$ dominates $\tilde{\mathbf{y}}$ in the sense of the multivariate **s**-increasing convex order if and only if $G_{\tilde{\mathbf{x}}}^{(k_1,\dots,k_N)}(\mathbf{c}) \ge G_{\tilde{\mathbf{y}}}^{(k_1,\dots,k_N)}(\mathbf{c})$ for all $c_i \in [\underline{x}_i, \bar{x}_i]$ if $k_i = s_i$ and $c_i = \underline{x}_i$ if $k_i = 1,\dots,s_i-1$, $i = 1,\dots,N$.

Mosler (1984) showed that stochastic dominance in terms of two special cases of the multivariate s-increasing concave order is related to multiplicative utilities. First, $\tilde{\mathbf{x}}$ dominates $\tilde{\mathbf{y}}$ in terms of the lower orthant order ($s = 1$) if and only if $\mathbb{E}[u(\tilde{\mathbf{x}})] \ge \mathbb{E}[u(\tilde{\mathbf{y}})]$ for all multiplicative utilities of the form $u(\mathbf{x}) = -\prod_{i=1}^{N}(-u_i(x_i))$, where $u_i(x_i) \le 0$ and $\frac{du_i(x_i)}{dx_i} \ge 0$ for all x_i, $i = 1,\dots,N$. Second, this dominance extends to the lower orthant concave order ($s = 2$) if each $u_i(x_i)$ is also concave. Theorem 2.5.7 extends these results to the multivariate **s**-increasing concave and convex orders for any **s**, showing that this order corresponds to the preferences of decision makers having utility functions consistent with mutual utility independence (Keeney and Raiffa 1976).

Theorem 2.5.7 *Let* $\tilde{\mathbf{x}}$ *and* $\tilde{\mathbf{y}}$ *be random vectors with support contained in* $[\underline{\mathbf{x}}, \bar{\mathbf{x}}]$, $-\infty < \underline{\mathbf{x}} < \bar{\mathbf{x}} < \infty$. *Then*

(1) $\tilde{\mathbf{x}}$ *dominates* $\tilde{\mathbf{y}}$ *in the sense of the multivariate* **s**-*increasing concave order if and only if*

$$(-1)^N \mathbb{E}\left[\prod_{i=1}^N u_i(\tilde{x}_i)\right] \leq (-1)^N \mathbb{E}\left[\prod_{i=1}^N u_i(\tilde{y}_i)\right]$$

for all $u_i \leq 0$, $u_i \in \underline{\mathbb{U}}^1_{s_i}$, $i = 1, \ldots, N$.

(2) $\tilde{\mathbf{x}}$ *dominates* $\tilde{\mathbf{y}}$ *in the sense of the multivariate* **s**-*increasing convex order if and only if*

$$\mathbb{E}\left[\prod_{i=1}^N u_i(\tilde{x}_i)\right] \geq \mathbb{E}\left[\prod_{i=1}^N u_i(\tilde{y}_i)\right]$$

for all $u_i \geq 0$, $u_i \in \overline{\mathbb{U}}^1_{s_i}$, $i = 1, \ldots, N$.

Proof For (1), suppose that $\tilde{\mathbf{x}}$ dominates $\tilde{\mathbf{y}}$ in the sense of the multivariate **s**-increasing concave order, and let

$$v(\mathbf{x}) = -\prod_{i=1}^N \big(-u_i(x_i)\big).$$

If $u_i \leq 0$ and $u_i \in \underline{\mathbb{U}}^1_{s_i}$, $i = 1, \ldots, N$, then $v \in \mathbb{U}^{\mathrm{N}}_{\mathbf{s}\text{-icv}}$. Therefore, $\mathbb{E}[v(\tilde{\mathbf{x}})] \geq \mathbb{E}[v(\tilde{\mathbf{y}})]$, so that

$$(-1)^N \mathbb{E}\left[\prod_{i=1}^N u_i(\tilde{x}_i)\right] \leq (-1)^N \mathbb{E}\left[\prod_{i=1}^N u_i(\tilde{y}_i)\right].$$

For the converse, suppose that

$$(-1)^N \mathbb{E}\left[\prod_{i=1}^N u_i(\tilde{x}_i)\right] \leq (-1)^N \mathbb{E}\left[\prod_{i=1}^N u_i(\tilde{y}_i)\right]$$

for all $u_i \leq 0$, $u_i \in \underline{\mathbb{U}}^1_{s_i}$, $i = 1, \ldots, N$. For $i = 1, \ldots, N$ and $k = 1, \ldots, s_i - 1$, let $u_i(x_i) = -(c_i - x_i)_+^{k_i+1}$ with $c_i = \bar{x}_i$ if $k_i < s_i$ and $c_i \in [\underline{x}, \bar{x}]$ if $k_i = s_i$. Thus, $u_i \leq 0$ and $u_i \in \underline{\mathbb{U}}^1_{s_i}$ if $k_i < s_i$. For $k_i = s_i$, u_i belongs to the closure of $\underline{\mathbb{U}}^1_{s_i}$ (i.e., there exists a sequence of functions $v_j \in \underline{\mathbb{U}}^1_{s_i}$, $j = 1, 2, \ldots$ with $v_j \to u_i$). Thus,

$$\mathbb{E}\left[\prod_{i=1}^N (c_i - \tilde{x}_i)_+^{k_i-1}\right] \leq \mathbb{E}\left[\prod_{i=1}^N (c_i - \tilde{y}_i)_+^{k_i-1}\right],$$

and by Theorem 2.5.5(1), $\tilde{\mathbf{x}}$ dominates $\tilde{\mathbf{y}}$ in the sense of the multivariate **s**-increasing concave order. Statement (2) follows from (1) and the duality between the concave and convex orders, as in the proof of Theorem 2.5.5. □

We now compare our multivariate dominance with dominance for the multivariate **s**-increasing orders. There are some close similarities between the two approaches and some important differences. In terms of infinite-degree stochastic dominance, the two approaches are equivalent, because

$$\lim_{\min\{s_i\}\to\infty} \mathbb{U}^{\mathrm{N}}_{\mathbf{s}\text{-icv}} = \underline{\mathbb{U}}^{N}_{\infty}$$

and

$$\lim_{\min\{s_i\}\to\infty} \mathbb{U}^{\mathrm{N}}_{\mathbf{s}\text{-icx}} = \overline{\mathbb{U}}^{N}_{\infty}.$$

However, this equivalence does not hold for finite n and **s**.

For finite n, nth-degree concave (convex) stochastic dominance is stronger than the n-increasing concave (convex) order, while the s-increasing concave (convex) order is stronger than (sN)th-degree concave (convex) stochastic dominance. In other words, (sN)th-degree concave (convex) stochastic dominance is between the s- and (sN)-increasing concave (convex) orders.

At a very basic level, our multivariate stochastic dominance is a natural extension of standard univariate stochastic dominance in that both are based on a preference between combining good with bad and combining good with good and bad with bad. A preference for combining good with bad leads to multivariate concave dominance and the most common univariate dominance. The opposite preference leads to multivariate convex dominance and a risk-taking version of univariate dominance. The preference condition is easy for decision makers to understand and therefore easy to check. If the decision maker has a consistent preference one way or the other, this implies corresponding constraints on the utility function via $\underline{\mathbb{U}}^{N}_{\infty}$ and $\overline{\mathbb{U}}^{N}_{\infty}$, but the discussion about preferences does not require direct consideration of utility.

Dominance in the sense of the **s**-increasing orders cannot be related to a simple preference assumption, but it can be characterized in terms of integral conditions that are extensions of the integral conditions for standard univariate dominance. In contrast, our multivariate dominance admits no such integral conditions. From a practical standpoint, however, the integral conditions in (2.4) and (2.5) might be difficult to verify as N increases or $\sum_{i=1}^{N} s_i$ increases.

Of course, not all decision makers share the same preferences. Thus, the preferences of different decision makers can be consistent with different classes of utility functions and therefore with different definitions of dominance. The approach to multivariate stochastic dominance developed here is intuitively appealing and should fit the preferences of some decision makers. As such, it is a useful addition to the stochastic dominance toolbox.

2.6 Summary and Conclusions

The concept of stochastic dominance has been widely studied in the univariate case, and there is widespread agreement on an underlying stochastic order for such dominance. This standard order is consistent with a basic preference condition, a preference for combining good with bad, as opposed to combining good with good and bad with bad. Many multivariate stochastic orders have been studied. However, most lack sufficient connections with the standard univariate stochastic dominance order and are not based on an intuitive preference condition that is easy to explain to decision makers. We fill this gap by defining multivariate nth-degree concave stochastic dominance and nth-degree risk in a way that naturally extends the univariate case because it is consistent with the same basic preference assumption. As in the univariate case, multivariate infinite-degree stochastic dominance is equivalent to an exponential ordering. We also develop the notion of multivariate convex stochastic dominance, which is consistent with a preference for combining good with good and bad with bad, as opposed to combining good with bad.

After developing our notion of multivariate stochastic dominance, we present some results that are useful in applying our multivariate stochastic dominance relations to rank alternatives. We show that independent additive or multiplicative background risk does not change stochastic dominance orderings and show how stochastic dominance can be applied to the choice among several alternatives using elimination by mixtures. We consider multivariate infinite-degree stochastic dominance, which is equivalent to an exponential ordering, as in the univariate case, and discuss the ordering of multivariate normal distributions. Finally, we discuss the connection of our approach with one based on a family of multivariate orders having some similarities to the order we use.

Many situations involve multiple decision makers, and somewhat divergent preferences can make decision making challenging. Even if each member of the group assesses a utility function (a challenging task itself, particularly in a multiattribute setting), it would be surprising for all members of the group to have identical utilities. However, the preferences of group members might be somewhat similar, especially when they are making a decision for their company and not a personal decision. They most likely will agree on a preference for more of each attribute to less or can define the attributes in such a way as to guarantee that preference, so that first-order stochastic dominance is applicable. They might also agree that the company's situation makes it prudent to be risk averse and that in general, a preference for combining good with bad is reasonable. This implies that they all should be willing to use a utility function $u \in \underline{\mathbb{U}}_n^N$ for any $n > 1$ and therefore to use multiattribute concave stochastic dominance to eliminate some alternatives from consideration.

Making a decision in a multiattribute situation is likely to be a multistage process. Some alternatives might be eliminated using stochastic dominance; choice among other alternatives might require more careful preference assessments, with emphasis on particular tradeoffs. That in turn might lead to clarification of objectives and attributes and generation of new promising alternatives (Keeney 1992). The results of our paper can be useful in that kind of decision process.

Acknowledgements We thank the referee and Editor for many helpful comments. The financial support of the "Onderzoeksfonds K.U. Leuven" (GOA/07: Risk Modeling and Valuation of Insurance and Financial Cash Flows, with Applications to Pricing, Provisioning, and Solvency) is gratefully acknowledged by Michel Denuit. Ilia Tsetlin was supported in part by the INSEAD Alumni Fund.

Part II
Downside Risk

Chapter 3
Reliable Quantification and Efficient Estimation of Credit Risk

Jörn Dunkel and Stefan Weber

3.1 Portfolio Models

Risk management in practice involves two complementary tasks: the construction of accurate portfolio models (Credit Suisse Financial Products 1997; Gupton et al. 1997; Gordy 2000; Frey and McNeil 2003; McNeil et al. 2005; Frey et al. 2008), and the reliable quantification of the downside risk for these models (Artzner et al. 1999; Föllmer and Schied 2011; Frey and McNeil 2002; Tasche 2002; Weber 2006). The first task covers both the design and the calibration of models to available data. In domains where data are scarce, models need to be extrapolated based on an understanding of the underlying economic mechanisms. The second task, the definition of well-defined benchmarks, is crucial since applied risk management and financial regulation require simple summary statistics that correctly reflect the loss exposure (Artzner et al. 1999; Frey and McNeil 2002; Tasche 2002).

A broad class of credit portfolio models (Credit Suisse Financial Products 1997; Gupton et al. 1997) specify the total loss $L \geq 0$ over a fixed period (e.g., a day, month, or year) by

$$L = \sum_{i=1}^{m} v_i D_i. \tag{3.1}$$

Here, m is the number of portfolio positions (obligors), v_i is the partial monetary loss that occurs if the obligor i defaults within this period, and D_i is the random default variable taking values 0 ("no default") or 1 ("default"). Realistic models

J. Dunkel
Department of Applied Mathematics and Theoretical Physics, University of Cambridge, Wilberforce Road, Cambridge CB3 0WA, UK

S. Weber (✉)
Leibniz Universität Hannover, Institut für Mathematische Stochastik, Welfengarten 1, 30167 Hannover, Germany
e-mail: sweber@stochastik.uni-hannover.de

F. Biagini et al. (eds.), *Risk Measures and Attitudes*, EAA Series,
DOI 10.1007/978-1-4471-4926-2_3,

take into account that the default risk of different positions may be interdependent (Credit Suisse Financial Products 1997; Gupton et al. 1997; Gordy 2000; Frey and McNeil 2003). Underlying mechanisms include observable and hidden economic risk factors, global feedback effects and local interactions (Embrechts et al. 2002). A pragmatic and popular way for modeling dependencies uses a factor structure (Credit Suisse Financial Products 1997; Gupton et al. 1997; Frey and McNeil 2003; Kang and Shahabuddin 2005). Within this approach, default indicators are constructed as binary functions $D_i = \Theta(\boldsymbol{a}_i \cdot \boldsymbol{Z} - x_i) \in \{0, 1\}$ with Θ denoting the Heaviside function $[\Theta(z) := 0, z \leq 0;\ \Theta(z) := 1, z > 0]$ and fixed threshold parameters $(x_1, \ldots, x_m)$. The random vector $\boldsymbol{Z} = (Z_1, \ldots, Z_d)$ comprises common or individual risk factors whose joint distribution is specified as an ingredient of the model. Potential dependencies between portfolio positions are encoded through coupling parameters $\boldsymbol{a}_i = (a_{ij})_{j=1,\ldots,d}$, which have to be deduced from historical data. For realistic models, it is usually impossible to analytically evaluate the full loss distribution $\mathbb{P}[L \leq x]$, and numerical simulations must be employed. A naive computer experiment would first sample the random numbers (Z_j) under the model probability measure $\mathbb{P}$ and subsequently calculate D_i and L. By repeating this procedure several times, one can estimate specific values of the loss distribution function, the mean loss $\mathbb{E}[L]$, the variance, or other relevant quantities. Of particular interest with regard to risk estimation are quantities which characterize extreme events that cause large losses (Artzner et al. 1999; Glasserman et al. 2002; Föllmer and Schied 2011; Giesecke et al. 2008). The recent market turmoil has clearly demonstrated that such scenarios may have serious global consequences for the stability of the financial system and the real economy, but they typically occur with very low probability; i.e., reliable predictions require advanced MC simulation techniques (Glasserman 2004). The novel method reported here allows for an efficient estimation and sensible characterization of big-loss scenarios through convex risk measures. The approach is generically applicable whenever the loss variable L can be sampled from a given set of rules similar to those outlined above.

3.2 Risk Measures

The theoretical foundations for the systematic measurement of financial risks were laid almost a decade ago (Artzner et al. 1999; Föllmer and Schied 2002); the numerical implementation of well-defined risk quantification schemes is, however, still a work-in-progress (Glasserman et al. 2002; Kang and Shahabuddin 2005). Risk measures, as specified in Eqs. (3.2) or (3.4) below, define the monetary amount s_* that should be available to insure against potentially large losses. The value s_* is called "capital requirement" or "economic capital" and depends on both the underlying portfolio model and the adopted risk measure. A major responsibility of regulatory authorities consists in identifying appropriate standards for risk measurement that prevent improper management of financial risks. Below, we describe an efficient MC method for estimating the risk measure *Shortfall Risk* (SR). Unlike the current industry standard of risk assessment *Value-at-Risk* (VaR) (Jorion 2000; Glasserman et al. 2002), SR encourages diversification and is well suited for characterizing rare big-loss scenarios. The severe deficiencies of VaR become evident upon analyzing its definition: For a fixed loss level $\lambda \in (0, 1)$, VaR is defined by

$$\begin{aligned} \mathrm{VaR}_\lambda &:= \inf\{s \in \mathbb{R} \mid \mathbb{P}[L > s] \le \lambda\} \\ &= \inf\{s \in \mathbb{R} \mid \mathbb{E}[\Theta(L - s)] \le \lambda\}. \end{aligned} \tag{3.2}$$

Representing a quantile of the loss distribution, VaR provides the threshold value that is exceeded by the loss L only with a small probability λ, but it ignores the shape of the loss distribution beyond the threshold. Very large losses are systematically underestimated by VaR. Consider e.g. a portfolio with loss distribution

$$L = \begin{cases} 0, & \text{with probability } 99.9\ \% \text{ (no loss)}, \\ \$10^{10}, & \text{with probability } 0.1\ \% \text{ (big loss)}. \end{cases} \tag{3.3}$$

Adopting the customary value $\lambda = 0.01$, one finds in this case $\mathrm{VaR}_\lambda = 0$, i.e., according to this risk measure, the portfolio does not require any economic capital although there exists a considerable chance of losing billions of dollars.

The severe deficiencies of VaR can be fixed by replacing the Θ-function in Eq. (3.2) with a convex, increasing loss function $\ell \ge 0$, which leads to the definition of SR, see e.g. Chap. 4.9 in Föllmer and Schied (2011):

$$\mathrm{SR}_\lambda := \inf\{s \in \mathbb{R} \mid \mathbb{E}[\ell(L - s)] \le \lambda\}, \tag{3.4}$$

where now $\lambda > 0$. Typical examples are exponential or (piecewise) polynomial loss functions,

$$\ell_\beta(y) = \exp(y/\beta), \qquad \ell_{\alpha,\eta}(y) = \eta^{-1}(y/\alpha)^\eta \Theta(y), \tag{3.5}$$

with scale parameters $\alpha, \beta > 0$ and $\eta \ge 1$. The function ℓ determines how strongly large losses are penalized. In the case of example (3.3), exponential SR with $\lambda = 0.01$ and $\beta = 2$ demands a capital requirement $s_* = \mathrm{SR}_\lambda^\beta(L) \approx \$10^9$, reflecting the actual size of potentially large losses. In contrast to VaR, SR risk measures provide a flexible tool for regulatory authorities to devise good risk measurement schemes.

3.3 Shortfall-Risk & Importance Sampling

Equation (3.4) implies that SR is equal to the unique root s_* of the function

$$g(s) := \mathbb{E}[\ell(L - s) - \lambda] \tag{3.6}$$

(see e.g. Chap. 4.9 in Föllmer and Schied 2011).

For realistic portfolio models, the functional value g at a given argument s can only be estimated numerically. A naive algorithm would sample n random variables

L_k according to the rules of the model, cf. Eq. (3.1), and compute the simple estimator $\hat{g}_n(s) = n^{-1}\sum_{k=1}^{n}\hat{G}(L_k, s)$ where $\hat{G}(L_k, s) = \ell(L_k - s) - \lambda$. This procedure is often inefficient for practically relevant loss distributions, since the variance of $\hat{g}_n(s)$ can be large. Improved estimates can be obtained by importance sampling (IS), defined as follows: Assume that L is governed by the probability density $p(x)$, abbreviated by $L \sim p$. For another, possibly s-dependent probability density $x \mapsto f_s(x)$, we may rewrite [1]

$$g(s) = -\lambda + \int \mathrm{d}x\, f_s(x) \frac{p(x)}{f_s(x)} \ell(x - s). \tag{3.7}$$

Consequently, $g_n(s) = n^{-1}\sum_{k=1}^{n} G(L_k, s)$ with

$$G(L_k, s) := -\lambda + \frac{p(L_k)}{f_s(L_k)}\, \ell(L_k - s), \quad L_k \sim f_s, \tag{3.8}$$

is another estimator for $g(s)$. Compared with the naive estimator $\hat{g}_n$, the variance of g_n can be substantially reduced if the IS density f_s is chosen appropriately (Dunkel and Weber 2007). Hence, to estimate SR one could try to combine IS with conventional root finding schemes, e.g., by defining a recursive sequence $s_j = R[s_{j-1}, \ldots, s_1; g(s_{j-1}), \ldots]$ using the secant method (Press et al. 2002). However, this approach suffers from drawbacks: Firstly, accurate estimates $g_n(s_j)$ of $g(s_j)$ at each point of the sequence $\{s_j\}$ are required which can be computationally expensive. Secondly, cross-averaging of errors for different values of s is not exploited. The algorithm below resolves these problems and yields a direct estimate of the SR value s_* by combining importance sampling with a stochastic root-finding scheme (Ruppert 1988, 1991; Polyak and Juditsky 1992).

3.4 Stochastic Root-Finding Algorithm

We focus here only on those aspects that are relevant for the practical implementation; a theoretical analysis can be found in Dunkel and Weber (2010). The proposed algorithm consists of the following steps:

1. Choose a fixed interval $[a, b]$ that contains the root s_*. Fix an initial value $s_1 \in [a, b]$ and constants $\gamma \in (\frac{1}{2}, 1]$ and $c > 0$.
2. Sample L_n from the IS density f_{s_n} and calculate

$$s_{n+1} = \Pi\left\{s_n + \frac{c}{n^\gamma} G(L_n, s_n)\right\}, \tag{3.9}$$

   where Π denotes a projection on the interval $[a, b]$, i.e., $\Pi\{x\} := a$ if $x < a$, $\Pi\{x\} := x$ if $x \in [a, b]$, and $\Pi\{x\} := b$ if $x > b$.

[1] $f_s(x)$ is assumed to be non-zero if $p(x)\ell(x - s) > 0$.

The sequence s_n defined by (3.9) converges to the SR value s_* as $n \to \infty$. More precisely, one can prove that, if c is chosen large enough, so that $c > [-2g'(s_*)]^{-1}$, then the distribution of the rescaled quantity

$$S_n := \sqrt{n^\gamma}(s_n - s_*) \tag{3.10}$$

converges to a Gaussian normal distribution $\mathcal{N}(\mu_*, \Sigma_*^2)$ with mean $\mu_* = 0$ and constant variance

$$\Sigma_*^2 = c^2\sigma^2(s_*)\begin{cases}[2c\,g'(s_*)]^{-1}, & \gamma \in (\frac{1}{2}, 1),\\ [2c\,g'(s_*)+1]^{-1}, & \gamma = 1,\end{cases} \tag{3.11}$$

where $\sigma^2(s)$ is the variance of the random variable $G(L, s)$ defined in (3.8). Equation (3.10) shows that γ determines the rate of convergence of the algorithm to s_*. The asymptotic variance in (3.11) can be improved by applying IS techniques that reduce the variance of $\sigma^2(s)$. This feature is particularly important when dealing with realistic loss distributions. Numerical values for the a priori unknown quantities $\sigma^2(s_*)$ and $g'(s_*)$ can be obtained from previously stored simulation data $\{(s_i, L_i, p(L_i), f_{s_i}(L_i))\}$ by using the numerically obtained root s_* to evaluate the estimators

$$\sigma_n^2(s_*) = \frac{1}{\rho n}\sum_{i=n(1-\rho)}^{n} G(L_i, s_i)^2, \quad \rho \in (0, 1), \tag{3.12}$$

$$g'_{n,\epsilon}(s_*) = \frac{1}{\epsilon n}\sum_{i=1}^{n}\left[\frac{p(L_i)}{f_{s_i}(L_i)}\ell\big(L_i - (s_* + \epsilon)\big) - \lambda\right] \tag{3.13}$$

for a sufficiently small $\epsilon > 0$. Estimates of $\sigma^2(s_*)$ and $g'(s_*)$ can be used for the construction of confidence intervals for s_*.

Variance reduction is not only important to decrease the asymptotic variance in (3.11), but also for improving the finite sample properties of the algorithm. S_n shows quasi-Gaussian behavior for much smaller values of n if IS is applied. At the same time, the estimators in (3.12) and (3.13) perform considerably better. The optimal choice of the constant c, which minimizes the variance in (3.11), is not known a priori. In practice, the optimal asymptotic variance can thus hardly be achieved. A solution to this problem is to average the estimator s_n given by (3.9) over the last $\rho \times n$ sampling steps, i.e., to return the estimator

$$\bar{s}_n = \frac{1}{\rho n}\sum_{i=n(1-\rho)}^{n} s_i, \quad \rho \in (0, 1). \tag{3.14}$$

In this case, one can show that, for $\gamma \in (\frac{1}{2}, 1)$ and $c > [-2g'(s_*)]^{-1}$, the distribution of the rescaled quantity $\bar{S}_n := \sqrt{\rho n}\,(\bar{s}_n - s_*)$ converges to the Gaussian distribution $\mathcal{N}(0, \sigma^2(s_*)/[g'(s_*)]^2)$ as $n \to \infty$. Apart from a factor $1/\rho$, the asymptotic variance then corresponds to the optimal choice for c in (3.11) in the case of an optimal convergence rate $\gamma = 1$.

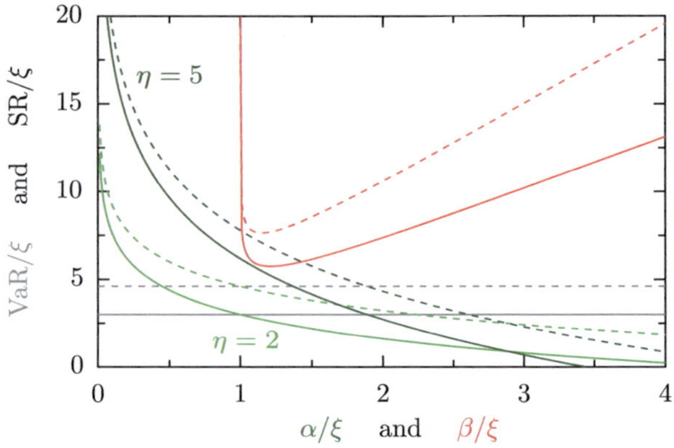

Fig. 3.1 Comparison of risk measures for a light-tailed exponential loss distribution (3.15): VaR_λ (*gray*), exponential $\mathrm{SR}_\lambda^\beta$ (*red*), and polynomial $\mathrm{SR}_\lambda^{\alpha,\eta}$ (*green*) in units of the mean loss ξ for levels $\lambda = 0.05$ (*solid*) and $\lambda = 0.01$ (*dashed*) plotted as functions of the rescaled parameters β/ξ and α/ξ, respectively

3.5 Applications

Due to the generic definition of the sequences s_n and $\bar{s}_n$, the above scheme is applicable to a wide range of portfolio models and can be combined with various model-specific variance reduction techniques (Glasserman 2004). To explicitly demonstrate the efficiency of the algorithms and to further illustrate the advantages of SR compared with VaR, we study two generic, stylized scenarios: a light-tailed exponential loss distribution with density

$$p(x) := \mathrm{d}\mathbb{P}[L < x]/\mathrm{d}x = \xi^{-1}\exp(-x/\xi)\Theta(x) \tag{3.15}$$

and a heavy-tailed power law distribution with density

$$p_\kappa(x) = \frac{(\kappa-1)[(\kappa-2)\xi]^{\kappa-1}}{[x+(\kappa-2)\xi]^\kappa}\Theta(x), \quad \kappa > 2, \tag{3.16}$$

where $\xi > 0$, respectively. In both cases, the mean loss is given by $\mathbb{E}[L] = \xi$, but ruinous losses are more likely to occur under the power law distribution (3.16).

For exponentially distributed losses and loss functions (3.5), the risk measures VaR and SR can be calculated analytically as

$$\begin{aligned} \mathrm{VaR}_\lambda &= \xi\log(\lambda^{-1}), \\ \mathrm{SR}_\lambda^\beta &= \beta\log[\lambda^{-1}(1-\xi/\beta)^{-1}], \\ \mathrm{SR}_\lambda^{\alpha,\eta} &= \xi\log[\lambda^{-1}(\xi/\alpha)^\eta\Gamma(\eta)], \end{aligned} \tag{3.17}$$

with Γ denoting the Gamma function. Finite positive SR values are obtained for $\beta > \xi$ and $\alpha < \xi[\Gamma(\eta)/\lambda]^{1/\eta}$. Figure 3.1 compares the three risk measures (3.17) for two values for λ. We plot the risk measures in units of ξ as functions of the normalized scale parameters α/ξ and β/ξ. For the exponential loss distribution, the probability of large losses increases with its mean value ξ. Figure 3.1 illustrates not only the dependence on α or β, respectively, but also how the risk measures behave

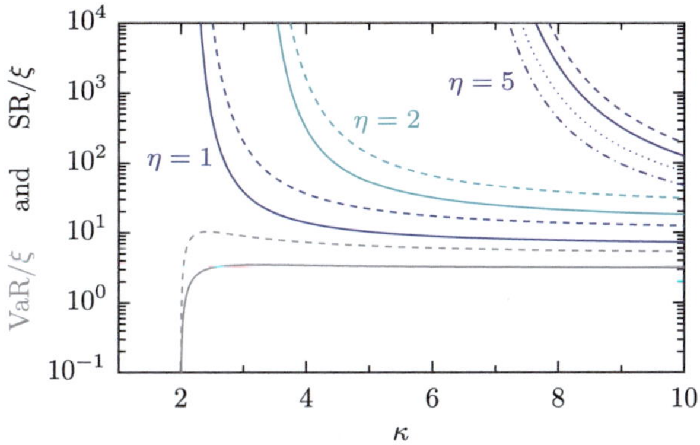

Fig. 3.2 VaR$_\lambda$ (*gray*) and polynomial SR$_\lambda^{\alpha,\eta}$ (*colored*) for the heavy-tailed distribution (3.16) plotted as a function of the exponent κ. *Solid* (*dashed*) *lines* correspond to levels $\lambda = 0.05$ (0.01), with $\alpha = 0.5$ in the case of SR. For SR with $\eta = 5$ (*violet*), additional curves with $\alpha = 1.0$, $\lambda = 0.05$ (*dash-dotted*) and $\alpha = 1.0$, $\lambda = 0.01$ (*dotted*) are shown. In the heavy-tail limit $\kappa \to 2$, VaR tends to zero and, thus, becomes inadequate for defining securities in this regime

as functions of the mean loss ξ. While VaR (gray) is proportional to ξ, polynomial SR (green) grows more than proportionally with ξ. Exponential SR (red), on the other hand, increases for small ξ less than proportionally but diverges rapidly as ξ approaches the parameter β. These specific characteristics must be taken into account by regulatory authorities and risk managers in order to devise and implement reasonable policies.

In the case of the heavy-tail distribution (3.16) exponential SR diverges, but VaR and polynomial SR with $1 \le \eta < \kappa - 1$ remain finite, yielding

$$\begin{aligned} \mathrm{VaR}_\lambda &= (\kappa - 2)\big(\lambda^{-1/(\kappa-1)} - 1\big)\xi, \\ \mathrm{SR}_\lambda^{\alpha,\eta} &= (2-\kappa)\xi + \left\{ \frac{[(\kappa-2)\xi]^{\kappa-1}}{\lambda\,\alpha^\eta\, C(\eta,\kappa)} \right\}^{1/(\kappa-1-\eta)}, \end{aligned} \tag{3.18}$$

where $C(\eta,\kappa) = \Gamma(\kappa-1)/[\Gamma(\eta)\,\Gamma(\kappa-1-\eta)]$. As evident from Eqs. (3.18) and Fig. 3.2, VaR (gray) vanishes in the heavy-tail limit $\kappa \to 2$, even though the tail risk is increased for smaller values of κ. By contrast, SR (colored) provides a reasonable risk measure for the whole parameter range.

We can use the analytic expressions (3.17) and (3.18) to verify the convergence behavior of the proposed algorithm. Figures 3.3 and 3.4 depict numerical results obtained from $N = 10^4$ sample runs for a fixed loss level $\lambda = 0.01$ and different values of n and γ (colors correspond to those in Figs. 3.1, 3.2). The diagrams show the sample mean values and variances of data sets $\{s_n^{(1)}, \dots, s_n^{(N)}\}$ and $\{\bar{s}_n^{(1)}, \dots, \bar{s}_n^{(N)}\}$, respectively. For each run (k) the initial value $s_1^{(k)}$ was randomly chosen from the search interval $[a,b] = [s_* - 5, s_* + 5]$ where s_* is the exact analytical value. One readily observes that in all examples the estimators converge to the exact values (dotted lines in Fig. 3.3). The convergence speed, however, depends on the underlying loss distribution and on the loss function. Generally, SR estimates based on

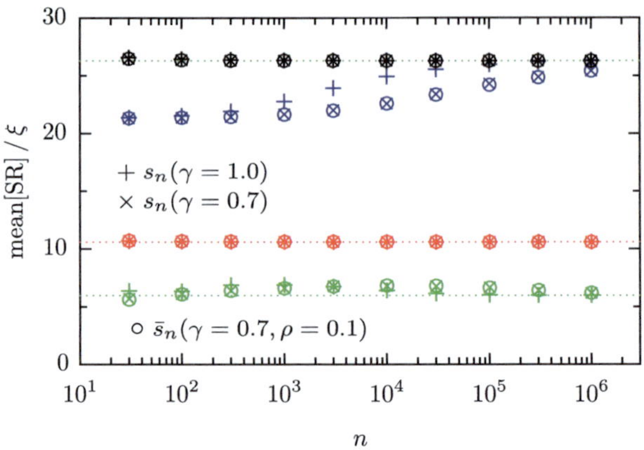

Fig. 3.3 Numerical SR estimates s_n and $\bar{s}_n$ as obtained from $N = 10^4$ simulation runs using $\lambda = 0.01$; the corresponding variances are depicted in Fig. 3.4. *Red/green symbols*: Exponential/polynomial SR for a light-tailed exponential loss distribution (3.15), using $c = 500$ and direct sampling. The estimators converge rapidly to the exact theoretical value (dotted, cf. Fig. 3.1) for exponential $\mathrm{SR}_\lambda^\beta$ ($\beta = 2\xi$; *red*), while the convergence is considerably slower for polynomial $\mathrm{SR}_\lambda^{\alpha,\eta}$ ($\alpha = 0.5\xi$, $\eta = 2$; *green*). *Blue/black symbols*: Polynomial $\mathrm{SR}_\lambda^{\alpha,\eta}$ estimated for the heavy-tail power-law distribution (3.16), using $c = 10^3$ and parameters $\alpha = 0.5\xi$, $\eta = 1$, $\kappa = 4$, cf. Fig. 3.2. Compared with direct sampling (*blue*), the importance sampling estimators (*black*) converge much faster

Fig. 3.4 Sample variances of the SR estimates from Fig. 3.3, using the same colors/symbols. Estimates can be considered as reliable when the variance of s_n ($+/\times$) or $\bar{s}_n$ ($\circ$) decreases with $n^{-\gamma}$ or n^{-1}, respectively. For the heavy-tail distribution (3.16), importance sampling (*black*) is much more efficient than direct sampling (*blue*)

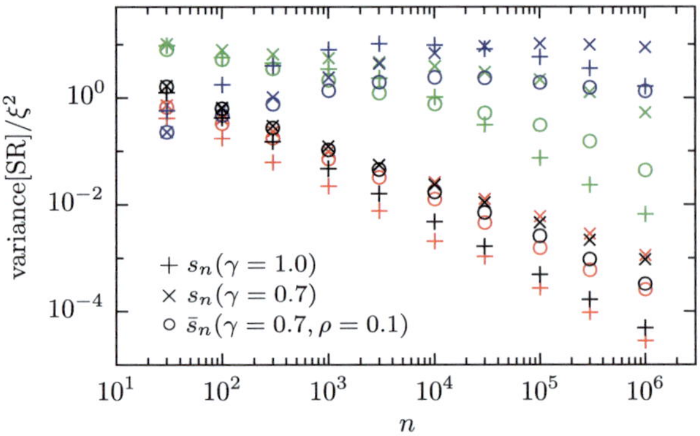

s_n or $\bar{s}_n$ can be considered reliable when the variance decreases with $n^{-\gamma}$ or n^{-1}, respectively, in accordance with Gaussian asymptotics for the rescaled quantities $S_n = \sqrt{n^\gamma}(s_n - s_*)$ and $\bar{S}_n = \sqrt{n}(\bar{s}_n - s_*)$. As evident from both Figs. 3.3 and 3.4, for the light-tailed exponential distribution $p(x) = \xi^{-1}\exp(-x/\xi)\Theta(x)$, the exponential SR estimators (red) converge very rapidly, while for polynomial SR (green), the convergence is slower but still acceptable even without importance sampling (i.e., if L_n is directly sampled from p).

By contrast, and not surprisingly, for the heavy-tail distribution (3.16), direct sampling (blue) of L_n from p_κ results in poor convergence behavior. In such cases, variance reduction techniques like IS (black) can significantly improve the performance. As guidance for future implementations, we outline the IS procedure in more detail: Instead of sampling losses from the original distribution p_κ, we consider the following "shifted" power law density

$$f_{\nu,s}(x) = \frac{(\nu-1)(\zeta+s)^{\nu-1}}{(x+\zeta)^\nu}\,\Theta(x-s), \tag{3.19}$$

where $\zeta > -s$ and $1 < \nu < 2(\kappa-\eta) - 1 =: \nu_+$. The latter condition ensures the finiteness of the second moment. We have to determine ν and ζ such that sampling from $f_{\nu,s}$ yields a better convergence to the correct SR value s_*. To this end, we note that the likelihood ratio $h(x) := p_\kappa(x)/f_{\nu,s}(x)$ takes its maximum at $x_- = s$, representing the effective lower integral boundary in Eq. (3.7), if

$$\zeta \ge \big[(\kappa-2)\nu\xi - (\kappa-\nu)s\big]/\kappa. \tag{3.20}$$

We fulfill this condition by fixing $\zeta = \nu(\xi+s)$. One then finds that variance reduction, corresponding to $h(x) < 1$ for $x \ge s$, is achieved if

$$\nu > \nu_*(s) := -\frac{2\kappa_1 s(\kappa_2\xi)^\kappa + \kappa_2\xi(s+\kappa_2\xi)^\kappa}{\kappa_1\xi(\kappa_2\xi)^\kappa - \kappa_2\xi(s+\kappa_2\xi)^\kappa}, \tag{3.21}$$

where $\kappa_2 = \kappa - 2, \kappa_1 = \kappa - 1$, and $\nu_* \to 1$ as $s \to \infty$. Accordingly, we sample $L_n \sim p_\kappa$ if $\nu_*(s_n) > \nu_+$, and $L_n \sim f_{\nu,s_n}$ with $\nu = 0.5[\nu_*(s_n) + \nu_+]$ if $\nu_*(s_n) < \nu_+$. Intuitively, by sampling from f_{ν,s_n} losses beyond s_n become more likely, while simultaneously suppressing the tail if $\nu_+ > \kappa$. As evident from Figs. 3.3 and 3.4, both aspects contribute to a vastly improved convergence. Most importantly, however, this strategy can be extended to more general models without much difficulty, e.g., by combining the stochastic root finding scheme with standard variance reduction techniques (Glasserman 2004) for the factor variables **Z** in Eq. (3.1).

3.6 Summary

Financial risk measures have been studied systematically for almost a decade (Artzner et al. 1999; Gordy 2000; Föllmer and Schied 2002; Weber 2006; McNeil et al. 2005). The financial industry, however, is still almost exclusively relying on the deficient risk measure Value-at-Risk (Glasserman et al. 2002; Jorion 2000), or even less sophisticated methodologies. The recent financial turmoil leaves little doubt about the importance of adequate risk quantification schemes. The above discussion clarifies how well-defined, tail-sensitive shortfall risk measures can be efficiently evaluated by combining stochastic root-approximation algorithms with variance reduction techniques. These tools can provide a basis for more sensible risk management policies and, thus, help to prevent future crises.

Acknowledgement The authors would like to thank Thomas Knispel for helpful remarks.

Chapter 4
Diffusion-Based Models for Financial Markets Without Martingale Measures

Claudio Fontana and Wolfgang J. Runggaldier

Keywords Arbitrage · Hedging · Contingent claim valuation · Market price of risk · Martingale deflator · Growth-optimal portfolio · Numéraire portfolio · Market completeness · Utility indifference valuation · Benchmark approach

4.1 Introduction

The concepts of *Equivalent (Local) Martingale Measure* (E(L)MM), *no-arbitrage*, and *risk-neutral pricing* can be rightfully considered as the cornerstones of modern mathematical finance. It seems to be almost folklore that such concepts can be regarded as mutually equivalent. In fact, most practical applications in quantitative finance are directly formulated under suitable assumptions which ensure that those concepts are indeed equivalent.

In recent years, maybe due to the dramatic turbulences raging over financial markets, an increasing attention has been paid to models that allow for financial market anomalies. More specifically, several authors have studied market models where stock price bubbles may occur (see e.g. Cox and Hobson 2005; Heston et al. 2007; Hulley 2010; Jarrow et al. 2007, 2010). It has been shown that bubble phenomena are consistent with the classical no-arbitrage theory based on the notion of *No Free Lunch with Vanishing Risk* (NFLVR), as developed in Delbaen and Schachermayer (1994) and Delbaen and Schachermayer (2006). However, in the presence of a bubble, discounted prices of risky assets are, under a risk-neutral measure, *strict* local martingales, i.e. local martingales which are not true martingales. This fact already implies that several well-known and classical results (for instance the *put–call parity* relation, see e.g. Cox and Hobson 2005) of mathematical finance do not hold anymore and must be modified accordingly.

C. Fontana (✉)
INRIA Paris-Rocquencourt, Domaine de Voluceau, Rocquencourt, BP 105, Le Chesnay Cedex 78153, France
e-mail: claudio.fontana@inria.fr

W.J. Runggaldier
University of Padova, Department of Mathematics, via Trieste 63, 35121 Padova, Italy
e-mail: runggal@math.unipd.it

F. Biagini et al. (eds.), *Risk Measures and Attitudes*, EAA Series,
DOI 10.1007/978-1-4471-4926-2_4,

A decisive step towards enlarging the scope of financial models has been represented by the study of models that do not fit at all into the classical no-arbitrage theory based on (NFLVR). Indeed, several authors (see e.g. Christensen and Larsen 2007; Delbaen and Schachermayer 1995a; Hulley 2010; Karatzas and Kardaras 2007; Loewenstein and Willard 2000) have studied instances where an ELMM may fail to exist. More specifically, financial models that do not admit an ELMM appear in the context of *Stochastic Portfolio Theory* (see Fernholz and Karatzas 2009 for a recent overview) and in the *Benchmark Approach* (see the monograph Platen and Heath 2006 for a detailed account). In the absence of a well-defined ELMM, many of the classical results of mathematical finance seem to break down, and one is led to ask whether there is still a meaningful way to proceed in order to solve the fundamental problems of portfolio optimisation and contingent claim valuation. It is then a remarkable result that a satisfactory theory can be developed even in the absence of an ELMM, especially in the case of a complete financial market model, as we are going to illustrate.

The present paper aims at carefully analysing a general class of diffusion-based financial models, without relying on the existence of an ELMM. More specifically, we discuss several notions of no-arbitrage that are weaker than the traditional (NFLVR) condition, and we study necessary and sufficient conditions for their validity. We show that the financial market may still be viable, in the sense that strong forms of arbitrage are banned from the market, even in the absence of an ELMM. In particular, it turns out that the viability of the financial market is fundamentally linked to a square-integrability property of the *market price of risk* process. Some of the results that we are going to present have already been obtained, also in more general settings (see e.g. Christensen and Larsen 2007; Chap. 4 of Fontana 2012; Hulley and Schweizer 2010; Karatzas and Kardaras 2007; Kardaras 2012, 2010). However, by exploiting the Itô-process structure, we are able to provide simple and transparent proofs, highlighting the key ideas behind the general theory. We also discuss the connections to the *Growth-Optimal Portfolio* (GOP), which is shown to be the unique portfolio possessing the *numéraire* property. In similar diffusion-based settings, related works that study the question of market viability in the absence of an ELMM include Fernholz and Karatzas (2009), Galesso and Runggaldier (2010), Heston et al. (2007), Loewenstein and Willard (2000), Londono (2004), Platen (2002) and Ruf (2012).

Besides studying the question of market viability, a major focus of this paper is on the valuation and hedging of contingent claims in the absence of an ELMM. In particular, we argue that the concept of *market completeness*, namely the capability to replicate every contingent claim, must be kept distinct from the existence of an ELMM. Indeed, we prove that the financial market may be viable and complete regardless of the existence of an ELMM. We then show that, in the context of a complete financial market, there is a unique natural candidate for the price of an arbitrary contingent claim, given by its GOP-discounted expected value under the original (*real-world*) probability measure. To this effect, we revisit some ideas originally appeared in the context of the *Benchmark Approach*, providing more careful proofs and extending some previous results.

The paper is structured as follows. Section 4.2 introduces the general setting, which consists of a class of Itô-process models satisfying minimal technical conditions. We introduce a basic standing assumption, and we carefully describe the set of admissible trading strategies. The question of whether (properly defined) arbitrage opportunities do exist or not is dealt with in Sect. 4.3. In particular, we explore the notions of *increasing profit* and *arbitrage of the first kind*, giving necessary and sufficient conditions for their absence from the financial market. In turn, this leads to the introduction of the concept of *martingale deflators*, which can be regarded as weaker counterparts to the traditional (density processes of) martingale measures. Section 4.4 proves the existence of a unique *Growth-Optimal* strategy, which admits an explicit characterization and also generates the *numéraire portfolio*. In turn, the latter is shown to be the reciprocal of a martingale deflator, thus linking the numéraire portfolio to the no-arbitrage criteria discussed in Sect. 4.3. Section 4.5 starts with the hedging and valuation of contingent claims, showing that the financial market may be complete even in the absence of an ELMM. Section 4.6 deals with contingent claim valuation according to three alternative approaches: *real-world pricing*, *upper-hedging pricing* and *utility indifference valuation*. In the particular case of a complete market, we show that they yield the same valuation formula. Section 4.7 concludes by pointing out possible extensions and further developments.

4.2 The General Setting

Let $(\Omega, \mathcal{F}, P)$ be a complete probability space. For a fixed time horizon $T \in (0, \infty)$, let $\mathbb{F} = (\mathcal{F}_t)_{0 \leq t \leq T}$ be a filtration on $(\Omega, \mathcal{F}, P)$ satisfying the usual conditions of right-continuity and completeness. Let $W = (W_t)_{0 \leq t \leq T}$ be an $\mathbb{R}^d$-valued Brownian motion on the filtered probability space $(\Omega, \mathcal{F}, \mathbb{F}, P)$. To allow for greater generality, we do not assume from the beginning that $\mathbb{F} = \mathbb{F}^W$, meaning that the filtration $\mathbb{F}$ may be strictly larger than the P-augmented Brownian filtration $\mathbb{F}^W$. Also, the initial σ-field $\mathcal{F}_0$ may be strictly larger than the trivial σ-field.

We consider a *financial market* composed of $N + 1$ securities $S^0, S^1, \ldots, S^N$, with $N \leq d$. As usual, we let S^0 represent a locally riskless asset, which we name *savings account*, and we define the process $S^0 = (S^0_t)_{0 \leq t \leq T}$ as follows:

$$S^0_t := \exp\left(\int_0^t r_u \, du\right) \quad \text{for } t \in [0, T], \tag{4.1}$$

where the *interest rate* process $r = (r_t)_{0 \leq t \leq T}$ is a real-valued progressively measurable process such that $\int_0^T |r_t| \, dt < \infty$ P-a.s. The remaining assets S^i, for $i = 1, \ldots, N$, are supposed to be *risky assets*. For every $i = 1, \ldots, N$, the process $S^i = (S^i_t)_{0 \leq t \leq T}$ is given by the solution to the following SDE:

$$dS^i_t = S^i_t \mu^i_t \, dt + \sum_{j=1}^d S^i_t \sigma^{i,j}_t \, dW^j_t, \qquad S^i_0 = s^i, \tag{4.2}$$

where:

(i) $s^i \in (0, \infty)$ for all $i = 1, \dots, N$;
(ii) $\mu = (\mu_t)_{0 \le t \le T}$ is an $\mathbb{R}^N$-valued progressively measurable process satisfying $\sum_{i=1}^N \int_0^T |\mu_t^i| \, dt < \infty$ P-a.s.;
(iii) $\sigma = (\sigma_t)_{0 \le t \le T}$ is an $\mathbb{R}^{N \times d}$-valued progressively measurable process satisfying $\sum_{i=1}^N \sum_{j=1}^d \int_0^T (\sigma_t^{i,j})^2 \, dt < \infty$ P-a.s.

The SDE (4.2) admits the following explicit solution, for every $i = 1, \dots, N$ and $t \in [0, T]$:

$$S_t^i = s^i \exp\left(\int_0^t \left(\mu_u^i - \frac{1}{2} \sum_{j=1}^d (\sigma_u^{i,j})^2 \right) du + \sum_{j=1}^d \int_0^t \sigma_u^{i,j} \, dW_u^j \right). \tag{4.3}$$

Note that conditions (ii)–(iii) above represent minimal conditions in order to have a meaningful definition of the ordinary and stochastic integrals appearing in (4.3). Apart from these technical requirements, we leave the stochastic processes μ and σ fully general. For $i = 0, 1, \dots, N$, we denote by $\bar{S}^i = (\bar{S}_t^i)_{0 \le t \le T}$ the *discounted price* process of the ith asset, defined as $\bar{S}_t^i := S_t^i / S_t^0$ for $t \in [0, T]$.

Let us now introduce the following standing assumption, which we shall always assume to be satisfied without any further mention.

Assumption 4.2.1 For all $t \in [0, T]$, the $(N \times d)$-matrix σ_t has P-a.s. full rank.

Remark 4.2.2 From a financial perspective, Assumption 4.2.1 means that the financial market does not contain redundant assets, i.e. there does not exist a non-trivial linear combination of $(S^1, \dots, S^N)$ that is locally riskless, in the sense that its dynamics are not affected by the Brownian motion W. However, we want to point out that Assumption 4.2.1 is only used in the following for proving uniqueness properties of trading strategies and, hence, could also be relaxed.

In order to rigorously describe the activity of trading in the financial market, we now introduce the concepts of *trading strategy* and *discounted portfolio process*. In the following definition we only consider *self-financing* trading strategies that generate positive portfolio processes.

Definition 4.2.3

(a) An $\mathbb{R}^N$-valued progressively measurable process $\pi = (\pi_t)_{0 \le t \le T}$ is an *admissible trading strategy* if $\int_0^T \|\sigma_t' \pi_t\|^2 \, dt < \infty$ P-a.s. and $\int_0^T |\pi_t'(\mu_t - r_t \mathbf{1})| \, dt < \infty$ P-a.s., where $\mathbf{1} := (1, \dots, 1)' \in \mathbb{R}^N$. We denote by $\mathcal{A}$ the set of all admissible trading strategies.

(b) For any $(v, \pi) \in \mathbb{R}_+ \times \mathcal{A}$, the *discounted portfolio process* $\bar{V}^{v,\pi} = (\bar{V}_t^{v,\pi})_{0\leq t\leq T}$ is defined by

$$\bar{V}_t^{v,\pi} := v\mathcal{E}\left(\sum_{i=1}^{N} \int \pi^i \frac{d\bar{S}^i}{\bar{S}^i}\right)_t$$
$$= v\exp\left(\int_0^t \pi_u'(\mu_u - r_u\mathbf{1})\,du - \frac{1}{2}\int_0^t \|\sigma_u'\pi_u\|^2\,du + \int_0^t \pi_u'\sigma_u\,dW_u\right) \tag{4.4}$$

for all $t \in [0, T]$, where $\mathcal{E}(\cdot)$ denotes the stochastic exponential (see e.g. Revuz and Yor 1999, Sect. IV.3).

The integrability conditions in part (a) of Definition 4.2.3 ensure that both the ordinary and the stochastic integrals appearing in (4.4) are well defined. For all $i = 1, \ldots, N$ and $t \in [0, T]$, π_t^i represents the proportion of wealth invested in the ith risky asset S^i at time t. Consequently, $1-\pi_t'\mathbf{1}$ represents the proportion of wealth invested in the savings account S^0 at time t. Note that part (b) of Definition 4.2.3 corresponds to requiring the trading strategy π to be *self-financing*. Observe that Definition 4.2.3 implies that, for any $(v, \pi) \in \mathbb{R}_+ \times \mathcal{A}$, we have $V_t^{v,\pi} = v\,V_t^{1,\pi}$ for all $t \in [0, T]$. Due to this scaling property, we shall often let $v = 1$ without loss of generality, denoting $V^{\pi} := V^{1,\pi}$ for any $\pi \in \mathcal{A}$. By definition, the discounted portfolio process $\bar{V}^{\pi}$ satisfies the following dynamics:

$$d\bar{V}_t^{\pi} = \bar{V}_t^{\pi} \sum_{i=1}^{N} \pi_t^i \frac{d\bar{S}_t^i}{\bar{S}_t^i} = \bar{V}_t^{\pi}\pi_t'(\mu_t - r_t\mathbf{1})\,dt + \bar{V}_t^{\pi}\pi_t'\sigma_t\,dW_t. \tag{4.5}$$

Remark 4.2.4 The fact that admissible portfolio processes are uniformly bounded from below by zero excludes pathological *doubling strategies* (see e.g. Karatzas and Shreve 1998, Sect. 1.1.2). Moreover, an economic motivation for focusing on positive portfolios only is given by the fact that market participants have *limited liability* and, therefore, are not allowed to trade anymore if their total tradeable wealth reaches zero. See also Sect. 2 of Christensen and Larsen (2007), Sect. 6 of Platen (2011) and Sect. 10.3 of Platen and Heath (2006) for an amplification of the latter point.

4.3 No-Arbitrage Conditions and the Market Price of Risk

In order to ensure that the model introduced in the previous section represents a viable financial market, in a sense to be made precise (see Definition 4.3.10), we need to carefully answer the question of whether properly defined arbitrage opportunities are excluded. We start by giving the following definition.

Definition 4.3.1 A trading strategy $\pi \in \mathcal{A}$ is said to yield an *increasing profit* if the corresponding discounted portfolio process $\bar{V}^\pi = (\bar{V}_t^\pi)_{0\le t\le T}$ satisfies the following two conditions:

(a) $\bar{V}^\pi$ is P-a.s. increasing, in the sense that

$$P\big(\bar{V}_s^\pi \le \bar{V}_t^\pi \text{ for all } s, t \in [0, T] \text{ with } s \le t\big) = 1;$$

(b) $P(\bar{V}_T^\pi > 1) > 0$.

The notion of increasing profit represents the most glaring type of arbitrage opportunity, and, hence, it is of immediate interest to know whether it is allowed or not in the financial market. As a preliminary, the following lemma gives an equivalent characterization of the notion of increasing profit. We denote by ℓ the Lebesgue measure on $[0, T]$.

Lemma 4.3.2 *There exists an increasing profit if and only if there exists a trading strategy $\pi \in \mathcal{A}$ satisfying the following two conditions*:

(a) $\pi_t' \sigma_t = 0$ $P \otimes \ell$*-a.e.*;
(b) $\pi_t'(\mu_t - r_t\mathbf{1}) \neq 0$ *on some subset of* $\Omega \times [0, T]$ *of positive* $P \otimes \ell$*-measure.*

Proof Let $\pi \in \mathcal{A}$ be a trading strategy yielding an increasing profit. Due to Definition 4.3.1, the process $\bar{V}^\pi$ is P-a.s. increasing, hence of finite variation. Equation (4.5) then implies that the continuous local martingale $(\int_0^t \bar{V}_u^\pi \pi_u' \sigma_u \, dW_u)_{0\le t\le T}$ is also of finite variation. This fact in turn implies that $\pi_t' \sigma_t = 0$ $P \otimes \ell$-a.e. (see e.g. Karatzas and Shreve 1991, Sect. 1.5). Since $P(\bar{V}_T^\pi > 1) > 0$, we must have $\pi_t'(\mu_t - r_t\mathbf{1}) \neq 0$ on some subset of $\Omega \times [0, T]$ of non-zero $P \otimes \ell$-measure.

Conversely, let $\pi \in \mathcal{A}$ be a trading strategy satisfying conditions (a)–(b). Define then the process $\bar{\pi} = (\bar{\pi}_t)_{0\le t\le T}$ as follows, for $t \in [0, T]$:

$$\bar{\pi}_t := \operatorname{sign}\big(\pi_t'(\mu_t - r_t\mathbf{1})\big)\pi_t.$$

It is clear that $\bar{\pi} \in \mathcal{A}$ and $\bar{\pi}_t' \sigma_t = 0$ $P \otimes \ell$-a.e., and hence, due to (4.4), for all $t \in [0, T]$,

$$\bar{V}_t^{\bar{\pi}} = \exp\left(\int_0^t \bar{\pi}_u'(\mu_u - r_u\mathbf{1})\, du\right).$$

Furthermore, we have that $\bar{\pi}_t'(\mu_t - r_t\mathbf{1}) \ge 0$, with strict inequality holding on some subset of $\Omega \times [0, T]$ of non-zero $P \otimes \ell$-measure. This implies that the process $\bar{V}^{\bar{\pi}} = (\bar{V}_t^{\bar{\pi}})_{0\le t\le T}$ is P-a.s. increasing and satisfies $P(\bar{V}_T^{\bar{\pi}} > 1) > 0$, thus showing that $\bar{\pi}$ yields an increasing profit. □

Remark 4.3.3 According to Definition 3.9 in Karatzas and Kardaras (2007), a trading strategy satisfying conditions (a)–(b) of Lemma 4.3.2 is said to yield an *immediate arbitrage opportunity* (see Delbaen and Schachermayer 1995b and Sect. 4.3.2 of Fontana 2012 for a thorough analysis of the concept). In a general semimartingale setting, Proposition 3.10 of Karatzas and Kardaras (2007) extends our Lemma 4.3.2 and shows that the absence of (unbounded) increasing profits is equivalent to the absence of immediate arbitrage opportunities.

The following proposition gives a necessary and sufficient condition in order to exclude the existence of increasing profits.

Proposition 4.3.4 *There are no increasing profits if and only if there exists an $\mathbb{R}^d$-valued progressively measurable process $\gamma = (\gamma_t)_{0\le t\le T}$ such that the following condition holds*:

$$\mu_t - r_t\mathbf{1} = \sigma_t\gamma_t \quad P\otimes\ell\text{-}a.e. \tag{4.6}$$

Proof Suppose that there exists an $\mathbb{R}^d$-valued progressively measurable process $\gamma = (\gamma_t)_{0\le t\le T}$ such that condition (4.6) is satisfied and let $\pi \in \mathcal{A}$ be such that $\pi_t'\sigma_t = 0$ $P\otimes\ell$-a.e. Then we have:

$$\pi_t'(\mu_t - r_t\mathbf{1}) = \pi_t'\sigma_t\gamma_t = 0 \quad P\otimes\ell\text{-a.e.},$$

meaning that there cannot exist a trading strategy $\pi \in \mathcal{A}$ satisfying conditions (a)–(b) of Lemma 4.3.2. Due to the equivalence result of Lemma 4.3.2, this implies that there are no increasing profits.

Conversely, suppose that there exists no trading strategy in $\mathcal{A}$ yielding an increasing profit. Let us first introduce the following linear spaces, for every $t \in [0,T]$:

$$\mathcal{R}(\sigma_t) := \{\sigma_t y : y \in \mathbb{R}^d\}, \qquad \mathcal{K}(\sigma_t') := \{y \in \mathbb{R}^N : \sigma_t' y = 0\}.$$

Denote by $\Pi_{\mathcal{K}(\sigma_t')}$ the orthogonal projection on $\mathcal{K}(\sigma_t')$. As in Lemma 1.4.6 of Karatzas and Shreve (1998), we define the process $p = (p_t)_{0\le t\le T}$ by

$$p_t := \Pi_{\mathcal{K}(\sigma_t')}(\mu_t - r_t\mathbf{1}).$$

Define then the process $\hat{\pi} = (\hat{\pi}_t)_{0\le t\le T}$ by

$$\hat{\pi}_t := \begin{cases} \frac{p_t}{\|p_t\|} & \text{if } p_t \neq 0, \\ 0 & \text{if } p_t = 0. \end{cases}$$

Since the processes μ and r are progressively measurable, Corollary 1.4.5 of Karatzas and Shreve (1998) ensures that $\hat{\pi}$ is progressively measurable. Clearly, we have then $\hat{\pi} \in \mathcal{A}$, and, by construction, $\hat{\pi}$ satisfies condition (a) of Lemma 4.3.2. Since there are no increasing profits, Lemma 4.3.2 implies that the following identity holds $P\otimes\ell$-a.e.:

$$\|p_t\| = \frac{p_t'}{\|p_t\|}(\mu_t - r_t\mathbf{1})\mathbb{1}_{\{p_t\neq 0\}} = \hat{\pi}_t'(\mu_t - r_t\mathbf{1})\mathbb{1}_{\{p_t\neq 0\}} = 0, \tag{4.7}$$

where the first equality uses the fact that $\mu_t - r_t\mathbf{1} - p_t \in \mathcal{K}^{\perp}(\sigma_t')$ for all $t \in [0,T]$, with the superscript $\perp$ denoting the orthogonal complement. From (4.7) we have

$p_t = 0$ $P \otimes \ell$-a.e., meaning that $\mu_t - r_t \mathbf{1} \in \mathcal{K}^{\perp}(\sigma_t') = \mathcal{R}(\sigma_t)$ $P \otimes \ell$-a.e. This amounts to saying that we have

$$\mu_t - r_t \mathbf{1} = \sigma_t \gamma_t \quad P \otimes \ell\text{-a.e.}$$

for some $\gamma_t \in \mathbb{R}^d$. Taking care of the measurability issues, it can be shown that we can take $\gamma = (\gamma_t)_{0 \le t \le T}$ as a progressively measurable process (compare Karatzas and Shreve 1998, the proof of Theorem 1.4.2). □

Let us now introduce one of the crucial objects in our analysis: the *market price of risk* process.

Definition 4.3.5 The $\mathbb{R}^d$-valued progressively measurable *market price of risk* process $\theta = (\theta)_{0 \le t \le T}$ is defined as follows, for $t \in [0, T]$:

$$\theta_t := \sigma_t' \left(\sigma_t \sigma_t'\right)^{-1} (\mu_t - r_t \mathbf{1}).$$

The standing Assumption 4.2.1 ensures that the market price of risk process θ is well defined.[1] From a financial perspective, θ_t measures the excess return $(\mu_t - r_t \mathbf{1})$ of the risky assets (with respect to the savings account) in terms of their volatility.

Remark 4.3.6 (Absence of increasing profits) Note that, by definition, the market price of risk process θ satisfies condition (4.6). Proposition 4.3.4 then implies that, under the standing Assumption 4.2.1, there are no increasing profits. Note however that θ may not be the unique process satisfying condition (4.6).

Let us now introduce the following integrability condition on the market price of risk process.

Assumption 4.3.7 The market price of risk process $\theta = (\theta_t)_{0 \le t \le T}$ belongs to $L^2_{\text{loc}}(W)$, meaning that $\int_0^T \|\theta_t\|^2 \, dt < \infty$ P-a.s.

Remark 4.3.8 Let $\gamma = (\gamma_t)_{0 \le t \le T}$ be an $\mathbb{R}^d$-valued progressively measurable process satisfying condition (4.6). Letting $\mathcal{R}(\sigma_t') = \{\sigma_t' x : x \in \mathbb{R}^N\}$ and $\mathcal{R}^{\perp}(\sigma_t') = \mathcal{K}(\sigma_t) = \{x \in \mathbb{R}^d : \sigma_t x = 0\}$, we can obtain the orthogonal decomposition $\gamma_t = \Pi_{\mathcal{R}(\sigma_t')}(\gamma_t) + \Pi_{\mathcal{K}(\sigma_t)}(\gamma_t)$, for $t \in [0, T]$. Under Assumption 4.2.1, elementary linear algebra gives that $\Pi_{\mathcal{R}(\sigma_t')}(\gamma_t) = \sigma_t'(\sigma_t \sigma_t')^{-1} \sigma_t \gamma_t = \sigma_t'(\sigma_t \sigma_t')^{-1}(\mu_t - r_t \mathbf{1}) = \theta_t$, thus giving $\|\gamma_t\| = \|\theta_t\| + \|\Pi_{\mathcal{K}(\sigma_t)}(\gamma_t)\| \ge \|\theta_t\|$ for all $t \in [0, T]$. This implies that, as soon as there exists *some* $\mathbb{R}^d$-valued progressively measurable process γ satisfying (4.6) and such that $\gamma \in L^2_{\text{loc}}(W)$, then the market price of risk process θ satisfies

[1] It is worth pointing out that, if Assumption 4.2.1 does not hold but condition (4.6) is satisfied, i.e. we have $\mu_t - r_t \mathbf{1} \in \mathcal{R}(\sigma_t)$ $P \otimes \ell$-a.e., then the market price of risk process θ can still be defined by replacing $\sigma_t'(\sigma_t \sigma_t')^{-1}$ with the *Moore–Penrose pseudoinverse* of the matrix σ_t.

Assumption 4.3.7. In other words, the risk premium process θ introduced in Definition 4.3.5 is minimal among all progressively measurable processes γ which satisfy condition (4.6).

Many of our results will rely on the key relation existing between Assumption 4.3.7 and no-arbitrage, which has been first examined in Ansel and Stricker (1992) and Schweizer (1992) and also plays a crucial role in Delbaen and Schachermayer (1995b) and Levental and Skorohod (1995). We now introduce a fundamental local martingale associated to the market price of risk process θ. Let us define the process $\widehat{Z} = (\widehat{Z}_t)_{0\leq t\leq T}$ as follows, for all $t \in [0, T]$:

$$\widehat{Z}_t := \mathcal{E}\left(-\int \theta' \, dW\right)_t = \exp\left(-\sum_{j=1}^{d}\int_0^t \theta_u^j \, dW_u^j - \frac{1}{2}\sum_{j=1}^{d}\int_0^t (\theta_u^j)^2 \, du\right). \tag{4.8}$$

Note that Assumption 4.3.7 ensures that the stochastic integral $\int \theta' \, dW$ is well defined as a continuous local martingale. It is well known that $\widehat{Z} = (\widehat{Z}_t)_{0\leq t\leq T}$ is a strictly positive continuous local martingale with $\widehat{Z}_0 = 1$. Due to Fatou's lemma, the process $\widehat{Z}$ is also a supermartingale (see e.g. Karatzas and Shreve 1991, Problem 1.5.19), and, hence, we have $\mathbb{E}[\widehat{Z}_T] \leq \mathbb{E}[\widehat{Z}_0] = 1$. It is easy to show that the process $\widehat{Z}$ is a true martingale, and not only a local martingale, if and only if $E[\widehat{Z}_T] = E[\widehat{Z}_0] = 1$. However, it may happen that the process $\widehat{Z}$ is a *strict* local martingale, i.e. a local martingale which is not a true martingale. In any case, the following proposition shows the basic property of the process $\widehat{Z}$.

Proposition 4.3.9 *Suppose that Assumption* 4.3.7 *holds and let* $\widehat{Z} = (\widehat{Z}_t)_{0\leq t\leq T}$ *be defined as in* (4.8). *Then the following hold*:

(a) *for all* $i = 1, \ldots, N$, *the process* $\widehat{Z}\,\bar{S}^i = (\widehat{Z}_t\,\bar{S}_t^i)_{0\leq t\leq T}$ *is a local martingale*;
(b) *for every* $\pi \in \mathcal{A}$, *the process* $\widehat{Z}\,\bar{V}^\pi = (\widehat{Z}_t\,\bar{V}_t^\pi)_{0\leq t\leq T}$ *is a local martingale.*

Proof Part (a) follows from part (b) by taking $\pi \in \mathcal{A}$ with $\pi^i \equiv 1$ and $\pi^j \equiv 0$ for $j \neq i$, for any $i = 1, \ldots, N$. Hence, it suffices to prove part (b). Recalling Eq. (4.5), an application of the product rule gives

$$\begin{aligned} d\big(\widehat{Z}_t \bar{V}_t^\pi\big) &= \bar{V}_t^\pi \, d\widehat{Z}_t + \widehat{Z}_t \, d\bar{V}_t^\pi + d\langle \bar{V}^\pi, \widehat{Z}\rangle_t \\ &= -\bar{V}_t^\pi \widehat{Z}_t \theta_t' \, dW_t + \widehat{Z}_t \bar{V}_t^\pi \pi_t'(\mu_t - r_t \mathbf{1}) \, dt + \widehat{Z}_t \bar{V}_t^\pi \pi_t' \sigma_t \, dW_t \\ &\quad - \widehat{Z}_t \bar{V}_t^\pi \pi_t' \sigma_t \theta_t \, dt \\ &= \widehat{Z}_t \, \bar{V}_t^\pi \big(\pi_t' \sigma_t - \theta_t'\big) \, dW_t. \end{aligned} \tag{4.9}$$

Since $\sigma'\pi \in L^2_{\mathrm{loc}}(W)$ and $\theta \in L^2_{\mathrm{loc}}(W)$, this shows the local martingale property of $\widehat{Z}\bar{V}^{\pi}$. □

Under the standing Assumption 4.2.1, we have seen that the diffusion-based financial market described in Sect. 4.2 does not allow for increasing profits (see Remark 4.3.6). However, the concept of increasing profit represents an almost pathological notion of arbitrage opportunity. Hence, we would like to know whether weaker and more economically meaningful types of arbitrage opportunities can exist. To this effect, let us give the following definition, adapted from Kardaras (2012).

Definition 4.3.10 An $\mathcal{F}$-measurable non-negative random variable ξ is called an *arbitrage of the first kind* if $P(\xi > 0) > 0$ and, for all $v \in (0,\infty)$, there exists a trading strategy $\pi^v \in \mathcal{A}$ such that $\bar{V}_T^{v,\pi^v} \geq \xi$ P-a.s. We say that the financial market is *viable* if there are no arbitrages of the first kind.

The following proposition shows that the existence of an increasing profit implies the existence of an arbitrage of the first kind. Due to the Itô-process framework considered in this paper, we are able to provide a simple proof.

Proposition 4.3.11 *Let $\pi \in \mathcal{A}$ be a trading strategy yielding an increasing profit. Then there exists an arbitrage of the first kind.*

Proof Let $\pi \in \mathcal{A}$ yield an increasing profit and define $\xi := \bar{V}_T^{\pi} - 1$. Due to Definition 4.3.1, the random variable ξ satisfies $P(\xi \geq 0) = 1$ and $P(\xi > 0) > 0$. Then, for any $v \in [1,\infty)$, we have $\bar{V}_T^{v,\pi} = v\bar{V}_T^{\pi} > v\xi \geq \xi$ P-a.s. Furthermore for any $v \in (0,1)$, let us define $\pi_t^v := -\frac{\log(v)+\log(1-v)}{v}\pi_t$. Clearly, for any $v \in (0,1)$, the process $\pi^v = (\pi_t^v)_{0\leq t\leq T}$ satisfies $\pi^v \in \mathcal{A}$ and, due to Lemma 4.3.2, $(\pi_t^v)'\sigma_t = 0$ $P \otimes \ell$-a.e. We have then:

$$\begin{aligned}
\bar{V}_T^{v,\pi^v} &= v\exp\biggl(\int_0^T (\pi_t^v)'(\mu_t - r_t\mathbf{1})\,dt\biggr) \\
&= v\bigl(\bar{V}_T^{\pi}\bigr)^{-\frac{\log(v)+\log(1-v)}{v}} > \bar{V}_T^{\pi} - 1 = \xi \quad P\text{-a.s.},
\end{aligned}$$

where the second equality follows from the elementary identity $\exp(\alpha x) = (\exp x)^{\alpha}$, and the last inequality follows since $vx^{-\frac{\log(v)+\log(1-v)}{v}} > x - 1$ for $x \geq 1$ and for every $v \in (0,1)$. We have thus shown that, for every $v \in (0,\infty)$, there exists a trading strategy $\pi^v \in \mathcal{A}$ such that $\bar{V}_T^{v,\pi^v} \geq \xi$ P-a.s. □

Remark 4.3.12 As we shall see by means of a simple example after Corollary 4.3.19, there are instances of models where there are no increasing profits but there are arbitrages of the first kind, meaning that the absence of arbitrages of the first kind is a *strictly* stronger no-arbitrage-type condition than the absence of increasing profits. Furthermore, there exists a notion of arbitrage opportunity lying between the notion of increasing profit and that of arbitrage of the first kind, namely the notion of *strong arbitrage opportunity*, which consists of a trading strategy $\pi \in \mathcal{A}$ such that $\bar{V}_t^\pi \geq 1$ P-a.s. for all $t \in [0, T]$ and $P(\bar{V}_T^\pi > 1) > 0$. It can be shown that there are no strong arbitrage opportunities if and only if there are no increasing profits *and* the process $(\int_0^t \|\theta_u\|^2 du)_{0 \leq t \leq T}$ does not jump to infinity on $[0, T]$. For simplicity of presentation, we omit the details and refer instead the interested reader to Theorem 3.5 of Strasser (2005) (where the absence of strong arbitrage opportunities is denoted as condition NA_+) and Sect. 4.3.2 of Fontana (2012). We want to point out that the notion of *strong arbitrage opportunity* plays an important role in the context of the *benchmark approach*, see e.g. Sect. 6 of Platen (2011), Sect. 10.3 of Platen and Heath (2006) and Remark 4.3.9 of Fontana (2012).

We now proceed with the question of whether arbitrages of the first kind are allowed in our financial market model. To this effect, let us first give the following definition.

Definition 4.3.13 A *martingale deflator* is a real-valued non-negative adapted process $D = (D_t)_{0 \leq t \leq T}$ with $D_0 = 1$ and $D_T > 0$ P-a.s. and such that the process $D\bar{V}^\pi = (D_t \bar{V}_t^\pi)_{0 \leq t \leq T}$ is a local martingale for every $\pi \in \mathcal{A}$. We denote by $\mathcal{D}$ the set of all martingale deflators.

Remark 4.3.14 Let $D \in \mathcal{D}$. Then, taking $\pi \equiv 0$, Definition 4.3.13 implies that D is a non-negative local martingale and hence, due to Fatou's lemma, also a supermartingale. Since $D_T > 0$ P-a.s., the minimum principle for non-negative supermartingales (see e.g. Revuz and Yor 1999, Proposition II.3.4) implies that $P(D_t > 0, D_{t-} > 0 \text{ for all } t \in [0, T]) = 1$.

Note that part (b) of Proposition 4.3.9 implies that, as soon as Assumption 4.3.7 is satisfied, the process $\widehat{Z} = (\widehat{Z}_t)_{0 \leq t \leq T}$ introduced in (4.8) is a martingale deflator, in the sense of Definition 4.3.13. The following lemma describes the general structure of martingale deflators. Related results can also be found in Ansel and Stricker (1992, 1993b) and Schweizer (1995).

Lemma 4.3.15 *Let $D = (D_t)_{0 \leq t \leq T}$ be a martingale deflator. Then there exist an $\mathbb{R}^d$-valued progressively measurable process $\gamma = (\gamma_t)_{0 \leq t \leq T}$ in $L^2_{\mathrm{loc}}(W)$ satisfying condition* (4.6) *and a real-valued local martingale $N = (N_t)_{0 \leq t \leq T}$ with $N_0 = 0$, $\Delta N > -1$ P-a.s. and $\langle N, W^i \rangle \equiv 0$ for all $i = 1, \ldots, d$, such that the following hold, for all $t \in [0, T]$*:

$$D_t = \mathcal{E}\left(-\int \gamma \, dW + N\right)_t. \tag{4.10}$$

Proof Let us define the process $L := \int D_-^{-1} dD$. Due to Remark 4.3.14, the process D_-^{-1} is well defined and, being adapted and left-continuous, is also predictable and locally bounded. Since D is a local martingale, this implies that the process L is well defined as a local martingale null at 0 and we have $D = \mathcal{E}(L)$. The Kunita–Watanabe decomposition (see Ansel and Stricker 1993a, case 3) allows us to represent the local martingale L as follows:

$$L = -\int \gamma \, dW + N$$

for some $\mathbb{R}^d$-valued progressively measurable process $\gamma = (\gamma_t)_{0 \le t \le T}$ belonging to $L^2_{\mathrm{loc}}(W)$, i.e. satisfying $\int_0^T \|\gamma_t\|^2 \, dt < \infty$ P-a.s., and for some local martingale $N = (N_t)_{0 \le t \le T}$ with $N_0 = 0$ and $\langle N, W^i \rangle \equiv 0$ for all $i = 1, \ldots, d$. Furthermore, since $\{D > 0\} = \{\Delta L > -1\}$ and $\Delta L = \Delta N$, we have that $\Delta N > -1$ P-a.s. It remains to show that γ satisfies condition (4.6). Let $\pi \in \mathcal{A}$. Then, by using the product rule and recalling Eq. (4.5), we have:

$$\begin{aligned}
d\big(D\bar{V}^{\pi}\big)_t &= D_{t-} \, d\bar{V}^{\pi}_t + \bar{V}^{\pi}_t \, dD_t + d\big\langle D, \bar{V}^{\pi} \big\rangle_t \\
&= D_{t-} \bar{V}^{\pi}_t \pi'_t (\mu_t - r_t \mathbf{1}) \, dt + D_{t-} \bar{V}^{\pi}_t \pi'_t \sigma_t \, dW_t + \bar{V}^{\pi}_t D_{t-} \, dL_t \\
&\quad + D_{t-} \bar{V}^{\pi}_t \, d\left\langle L, \int \pi' \sigma \, dW \right\rangle_t \\
&= D_{t-} \bar{V}^{\pi}_t \pi'_t (\mu_t - r_t \mathbf{1}) \, dt + D_{t-} \bar{V}^{\pi}_t \pi'_t \sigma_t \, dW_t + \bar{V}^{\pi}_t D_{t-} \, dL_t \\
&\quad - D_{t-} \bar{V}^{\pi}_t \pi'_t \sigma_t \gamma_t \, dt \\
&= D_{t-} \bar{V}^{\pi}_t \pi'_t \sigma_t \, dW_t + \bar{V}^{\pi}_t D_{t-} \, dL_t \\
&\quad + D_{t-} \bar{V}^{\pi}_t \pi'_t (\mu_t - r_t \mathbf{1} - \sigma_t \gamma_t) \, dt.
\end{aligned} \tag{4.11}$$

Since $D \in \mathcal{D}$, the product $D\bar{V}^{\pi}$ is a local martingale for every $\pi \in \mathcal{A}$. This implies that the continuous finite-variation term in (4.11) must vanish. Since D_- and $\bar{V}^{\pi}$ are P-a.s. strictly positive and $\pi \in \mathcal{A}$ was arbitrary, this implies that condition (4.6) must hold. □

The following proposition shows that the existence of a martingale deflator is a sufficient condition for the absence of arbitrages of the first kind.

Proposition 4.3.16 *If $\mathcal{D} \neq \emptyset$, then there cannot exist arbitrages of the first kind.*

Proof Let $D \in \mathcal{D}$ and suppose that there exists a random variable ξ yielding an arbitrage of the first kind. Then, for every $n \in \mathbb{N}$, there exists a strategy $\pi^n \in \mathcal{A}$ such that $\bar{V}^{1/n,\pi^n}_T \ge \xi$ P-a.s. For every $n \in \mathbb{N}$, the process $D\bar{V}^{1/n,\pi^n} = (D_t \bar{V}^{1/n,\pi^n}_t)_{0 \le t \le T}$ is a positive local martingale and, hence, a supermartingale. So, for every $n \in \mathbb{N}$,

$$E[D_T \xi] \le E\big[D_T \bar{V}^{1/n,\pi^n}_T\big] \le E\big[D_0 \bar{V}^{1/n,\pi^n}_0\big] = \frac{1}{n}.$$

Letting $n \to \infty$ gives $E[D_T \xi] = 0$ and hence $D_T \xi = 0$ P-a.s. Since, due to Definition 4.3.13, we have $D_T > 0$ P-a.s., this implies that $\xi = 0$ P-a.s., which contradicts the assumption that ξ is an arbitrage of the first kind. □

It is worth pointing out that one can also prove a converse result to Proposition 4.3.16, showing that if there are no arbitrages of the first kind, then the set $\mathcal{D}$ is non-empty. In a general semimartingale setting, this has been recently shown in Kardaras (2012) (see also Sect. 4 of Fontana 2012 and Hulley and Schweizer 2010 in the context of continuous-path processes). Furthermore, Proposition 1 of Kardaras (2010) shows that the absence of arbitrages of the first kind is equivalent to the condition of *No Unbounded Profit with Bounded Risk* (NUPBR), formally defined as the condition that the set $\{\bar{V}_T^\pi : \pi \in \mathcal{A}\}$ be bounded in probability.[2] By relying on these facts, we can state the following theorem,[3] the second part of which follows from Proposition 4.19 of Karatzas and Kardaras (2007).

Theorem 4.3.17 *The following are equivalent*:

(a) $\mathcal{D} \neq \emptyset$;
(b) *there are no arbitrages of the first kind*;
(c) $\{\bar{V}_T^\pi : \pi \in \mathcal{A}\}$ *is bounded in probability, i.e. the* (*NUPBR*) *condition holds.*

Moreover, for every concave and strictly increasing utility function $U : [0, \infty) \to \mathbb{R}$, *the expected utility maximisation problem of finding an element* $\pi^* \in \mathcal{A}$ *such that*

$$E\big[U\big(\bar{V}_T^{\pi^*}\big)\big] = \sup_{\pi \in \mathcal{A}} E\big[U\big(\bar{V}_T^{\pi}\big)\big]$$

either does not have a solution or has infinitely many solutions when any of conditions (a)–(c) *fails.*

In view of the second part of the above theorem, the condition of absence of arbitrages of the first kind can be seen as the minimal no-arbitrage condition in order to be able to meaningfully solve portfolio optimisation problems.

Remark 4.3.18 We have defined the notion of *viability* for a financial market in terms of the absence of arbitrages of the first kind (see Definition 4.3.10). In Loewenstein and Willard (2000), a financial market is said to be viable if any agent with sufficiently regular preferences and with a positive initial endowment can construct an optimal portfolio. The last part of Theorem 4.3.17 gives a correspondence between these two notions of viability, since it shows that the absence of arbitrages of the first kind is the minimal no-arbitrage-type condition in order to being able to meaningfully solve portfolio optimisation problems.

---

[2]The (NUPBR) condition has been introduced under that name in Karatzas and Kardaras (2007). However, the condition that the set $\{\bar{V}_T^\pi : \pi \in \mathcal{A}\}$ be bounded in probability also plays a key role in the seminal work Delbaen and Schachermayer (1994), and its implications have been systematically studied in Kabanov (1997), where the same condition is denoted as "property BK".

[3]We want to remark that an analogous result has already been given in Theorem 2 of Loewenstein and Willard (2000) under the assumption of a complete financial market.

It is now straightforward to show that, as soon as Assumption 4.3.7 holds, the diffusion-based model introduced in Sect. 4.2 satisfies the equivalent conditions of Theorem 4.3.17. In fact, due to Proposition 4.3.9, the process $\widehat{Z}$ defined in (4.8) is a martingale deflator for the financial market $(S^0, S^1, \dots, S^N)$ as soon as Assumption 4.3.7 is satisfied, and, hence, due to Proposition 4.3.16, there are no arbitrages of the first kind. Conversely, suppose that there are no arbitrages of the first kind but Assumption 4.3.7 fails to hold. Then, due to Remark 4.3.8 together with Lemma 4.3.15, we have that $\mathcal{D} = \emptyset$. Theorem 4.3.17 then implies that there exist arbitrages of the first kind, thus leading to a contradiction. We have thus proved the following corollary.

Corollary 4.3.19 *The financial market* $(S^0, S^1, \dots, S^N)$ *is viable, i.e. it does not admit arbitrages of the first kind* (*see Definition* 4.3.10), *if and only if Assumption* 4.3.7 *holds.*

As we have seen in Proposition 4.3.11, if there exists an increasing profit, then there exist an arbitrage of the first kind. We now show that the absence of arbitrages of the first kind is a *strictly* stronger no-arbitrage-type condition than the absence of increasing profits by means of a simple example, which we adapt from Example 3.4 of Delbaen and Schachermayer (1995b). Let $N = d = 1$ and $r \equiv 0$, and let the real-valued process $S = (S_t)_{0 \le t \le T}$ be given as the solution to the following SDE:

$$dS_t = \frac{S_t}{\sqrt{t}}\,dt + S_t\,dW_t, \qquad S_0 = s \in (0, \infty).$$

Using the notation introduced in Sect. 4.2, we have $\mu_t = 1/\sqrt{t}$ for $t \in [0, T]$ and $\sigma \equiv 1$. Clearly, condition (4.6) is satisfied, since we trivially have $\mu_t = \sigma_t \theta_t$, where $\theta_t = 1/\sqrt{t}$ for $t \in [0, T]$. Proposition 4.3.4 then implies that there are no increasing profits. However, $\theta \notin L^2_{\text{loc}}(W)$, since $\int_0^t \theta_u^2\,du = \int_0^t \frac{1}{u}\,du = \infty$ for all $t \in [0, T]$. Corollary 4.3.19 then implies that there exist arbitrages of the first kind.[4]

We want to emphasise that, due to Theorem 4.3.17, the diffusion-based model introduced in Sect. 4.2 allows us to meaningfully consider portfolio optimisation problems as soon as Assumption 4.3.7 holds. However, nothing guarantees that an *Equivalent Local Martingale Measure* (ELMM) exists, as shown in the following classical example, already considered in Delbaen and Schachermayer (1995a), Hulley (2010) and Karatzas and Kardaras (2007). Other instances of models for which an ELMM does not exist arise in the context of *diverse* financial markets, see Chap. II of Fernholz and Karatzas (2009).

---

[4]More precisely, note that the process $(\int_0^t \theta_u^2\,du)_{0 \le t \le T} = (\int_0^t \frac{1}{u}\,du)_{0 \le t \le T}$ jumps to infinity instantaneously at $t = 0$. Hence, as explained in Remark 4.3.12, the model considered in the present example allows not only for arbitrages of the first kind, but also for strong arbitrage opportunities. Of course, there are instances where strong arbitrage opportunities are precluded, but still there exist arbitrages of the first kind. We refer the interested reader to Ball and Torous (1983) for an example of such a model, where the price of a risky asset is modelled as the exponential of a *Brownian bridge* (see also Loewenstein and Willard 2000, Example 3.1).

Example Let us suppose that $\mathbb{F} = \mathbb{F}^W$, where W is a standard Brownian motion ($d = 1$), and let $N = 1$. Assume that $S_t^0 \equiv 1$ for all $t \in [0, T]$ and that the real-valued process $S = (S_t)_{0 \le t \le T}$ is given by the solution to the following SDE:

$$dS_t = \frac{1}{S_t}\,dt + dW_t, \qquad S_0 = s \in (0, \infty). \tag{4.12}$$

It is well known that S is a Bessel process of dimension three (see e.g. Revuz and Yor 1999, Sect. XI.1). So, S_t is P-a.s. strictly positive and finite valued for all $t \in [0, T]$. The market price of risk process θ is given by $\theta_t = \sigma_t^{-1}\,\mu_t = \frac{1}{S_t}$ for $t \in [0, T]$. Since S is continuous, we clearly have $\int_0^T \theta_t^2\,dt < \infty$ P-a.s., meaning that Assumption 4.3.7 is satisfied. Hence, due to Corollary 4.3.19, there are no arbitrages of the first kind.

However, for this particular financial market model, there exists no ELMM. We prove this claim arguing by contradiction. Suppose that Q is an ELMM for S and denote by $Z^Q = (Z_t^Q)_{0 \le t \le T}$ its density process. Then, due to the martingale representation theorem (see Karatzas and Shreve 1991, Theorem 3.4.15 and Problem 3.4.16), we can represent Z^Q as follows:

$$Z_t^Q = \mathcal{E}\left(-\int \lambda\,dW\right)_t \qquad \text{for } t \in [0, T],$$

where $\lambda = (\lambda_t)_{0 \le t \le T}$ is a progressively measurable process with $\int_0^T \lambda_t^2\,dt < \infty$ P-a.s. Due to Girsanov's theorem, the process $W^Q = (W_t^Q)_{0 \le t \le T}$ defined by $W_t^Q := W_t + \int_0^t \lambda_u\,du$, for $t \in [0, T]$, is a Brownian motion under Q. Hence, the process S satisfies the following SDE under Q:

$$dS_t = \left(\frac{1}{S_t} - \lambda_t\right)dt + dW_t^Q, \qquad S_0 = s. \tag{4.13}$$

Since Q is an ELMM for S, the SDE (4.13) must have a zero drift term, i.e. it must be $\lambda_t = \frac{1}{S_t} = \theta_t$ for all $t \in [0, T]$. Then, a simple application of Itô's formula gives

$$Z_t^Q = \mathcal{E}\left(-\int \frac{1}{S}\,dW\right)_t = \exp\left(-\int_0^t \frac{1}{S_u}\,dW_u - \frac{1}{2}\int_0^t \frac{1}{S_u^2}\,du\right) = \frac{1}{S_t}.$$

However, since S is a Bessel process of dimension three, it is well known that the process $1/S = (1/S_t)_{0 \le t \le T}$ is a strict local martingale, i.e. it is a local martingale but not a true martingale (see e.g. Revuz and Yor 1999, Exercise XI.1.16). Clearly, this contradicts the fact that Q is a well-defined probability measure,[5] thus showing that there cannot exist an ELMM for S.

[5] Alternatively, one can show that the probability measures Q and P fail to be equivalent by arguing as follows. Let us define the stopping time $\tau := \inf\{t \in [0, T] : S_t = 0\}$. The process $S = (S_t)_{0 \le t \le T}$ is a Bessel process of dimension three under P, and, hence, we have $P(\tau < \infty) = 0$. However, since the process $S = (S_t)_{0 \le t \le T}$ is a Q-Brownian motion, we clearly have $Q(\tau < \infty) > 0$. This contradicts the assumption that Q and P are equivalent.

As the above example shows, Assumption 4.3.7 does not guarantee the existence of an ELMM for the financial market $(S^0, S^1, \dots, S^N)$. It is well known that, in the case of continuous-path processes, the existence of an ELMM is equivalent to the *No Free Lunch with Vanishing Risk* (NFLVR) no-arbitrage-type condition, see Delbaen and Schachermayer (1994) and Delbaen and Schachermayer (2006). Furthermore, the NFLVR condition holds if and only if both NUPBR and the classical *no-arbitrage* (NA) conditions hold (see Sect. 3 of Delbaen and Schachermayer 1994, Lemma 2.2 of Kabanov 1997 and Proposition 4.2 of Karatzas and Kardaras 2007), where, recalling that $\bar{V}_0^\pi = 1$, the NA condition precludes the existence of a trading strategy $\pi \in \mathcal{A}$ such that $P(\bar{V}_T^\pi \geq 1) = 1$ and $P(\bar{V}_T^\pi > 1) > 0$. This implies that, even if Assumption 4.3.7 holds, the classical NFLVR condition may fail to hold. However, due to Theorem 4.3.17, the financial market may still be viable.

Remark 4.3.20 (On the martingale property of $\widehat{Z}$) It is important to note that Assumption 4.3.7 does not suffice to ensure that $\widehat{Z}$ is a true martingale. Well-known sufficient conditions for this to hold include the Novikov and Kazamaki criteria, see e.g. Revuz and Yor (1999), Sect. VIII.1. If $\widehat{Z}$ is a true martingale, we have then $\mathbb{E}[\widehat{Z}_T] = 1$, and we can define a probability measure $\widehat{Q} \sim P$ by letting $\frac{d\widehat{Q}}{dP} := \widehat{Z}_T$. The martingale $\widehat{Z}$ represents then the *density process* of $\widehat{Q}$ with respect to P, i.e. $\widehat{Z}_t = \mathbb{E}[\frac{d\widehat{Q}}{dP} | \mathcal{F}_t]$ P-a.s. for all $t \in [0, T]$, and a process $M = (M_t)_{0 \leq t \leq T}$ is a local $\widehat{Q}$-martingale if and only if the process $\widehat{Z}M = (\widehat{Z}_t M_t)_{0 \leq t \leq T}$ is a local P-martingale. Due to Proposition 4.3.9(a), this implies that if $\mathbb{E}[\widehat{Z}_T] = 1$ then the process $\bar{S} := (\bar{S}^1, \dots, \bar{S}^N)'$ is a local $\widehat{Q}$-martingale or, in other words, the probability measure $\widehat{Q}$ is an ELMM. Girsanov's theorem then implies that the process $\widehat{W} = (\widehat{W}_t)_{0 \leq t \leq T}$ defined by $\widehat{W}_t := W_t + \int_0^t \theta_u \, du$ for $t \in [0, T]$ is a Brownian motion under $\widehat{Q}$. Since the dynamics of $S := (S^1, \dots, S^N)'$ in (4.2) can be rewritten as

$$dS_t = \operatorname{diag}(S_t)\mathbf{1}\, r_t \, dt + \operatorname{diag}(S_t)\sigma_t(\theta_t \, dt + dW_t), \qquad S_0 = s,$$

the process $\bar{S} := (\bar{S}^1, \dots, \bar{S}^N)'$ satisfies the following SDE under the measure $\widehat{Q}$:

$$d\bar{S}_t = \operatorname{diag}(\bar{S}_t)\sigma_t \, d\widehat{W}_t, \qquad \bar{S}_0 = s.$$

We want to point out that the process $\widehat{Z} = (\widehat{Z}_t)_{0 \leq t \leq T}$ represents the density process with respect to P of the *minimal martingale measure*, when the latter exists, see e.g. Hulley and Schweizer (2010). Again, we emphasise that in this paper we do not assume neither that $\mathbb{E}[\widehat{Z}_T] = 1$ nor that an ELMM exists.

We close this section with a simple technical result that turns out to be useful in the following.

Lemma 4.3.21 *Suppose that Assumption* 4.3.7 *holds. An* $\mathbb{R}^N$*-valued progressively measurable process* $\pi = (\pi_t)_{0 \leq t \leq T}$ *belongs to* $\mathcal{A}$ *if and only if* $\int_0^T \|\sigma_t' \pi_t\|^2 \, dt < \infty$ P*-a.s.*

Proof We only need to show that Assumption 4.3.7 and $\int_0^T \|\sigma_t'\pi_t\|^2\,dt < \infty$ P-a.s. together imply that $\int_0^T |\pi_t'(\mu_t - r_t\mathbf{1})|\,dt < \infty$ P-a.s. This follows easily from the Cauchy–Schwarz inequality:

$$\int_0^T \left|\pi_t'(\mu_t - r_t\mathbf{1})\right| dt = \int_0^T \left|\pi_t'\sigma_t\,\theta_t\right| dt$$
$$\le \left(\int_0^T \|\sigma_t'\pi_t\|^2\,dt\right)^{\frac{1}{2}}\left(\int_0^T \|\theta_t\|^2\,dt\right)^{\frac{1}{2}} < \infty \quad P\text{-a.s.} \qquad \square$$

4.4 The Growth-Optimal Portfolio and the Numéraire Portfolio

As we have seen in the last section, the diffusion-based model introduced in Sect. 4.2 can represent a viable financial market even if the traditional (NFLVR) no-arbitrage-type condition fails to hold or, equivalently, if an ELMM for $(S^0, S^1, \dots, S^N)$ fails to exist. Let us now consider an interesting portfolio optimisation problem, namely the problem of maximising the *growth rate*, formally defined as follows (compare Fernholz and Karatzas 2009; Platen 2006 and Platen and Heath 2006, Sect. 10.2).

Definition 4.4.1 For a trading strategy $\pi \in \mathcal{A}$, we call *growth rate* process the process $g^\pi = (g_t^\pi)_{0\le t\le T}$ appearing in the drift term of the SDE satisfied by the process $\log V^\pi = (\log V_t^\pi)_{0\le t\le T}$, i.e. the term g_t^π in the SDE

$$d\log V_t^\pi = g_t^\pi\,dt + \pi_t'\sigma_t\,dW_t. \tag{4.14}$$

A trading strategy $\pi^* \in \mathcal{A}$ (and the corresponding portfolio process V^{π^*}) is said to be *growth-optimal* if $g_t^{\pi^*} \ge g_t^\pi$ P-a.s. for all $t \in [0,T]$ for any trading strategy $\pi \in \mathcal{A}$.

The terminology "growth rate" is motivated by the fact that

$$\lim_{T\to\infty}\frac{1}{T}\left(\log V_T^\pi - \int_0^T g_t^\pi\,dt\right) = 0 \quad P\text{-a.s.}$$

under "controlled growth" of $a := \sigma\sigma'$, i.e. $\lim_{T\to\infty}(\frac{\log\log T}{T^2}\int_0^T a_t^{i,i}\,dt) = 0$ P-a.s. (see Fernholz and Karatzas 2009, Sect. 1). In the context of the general diffusion-based financial market described in Sect. 4.2, the following theorem gives an explicit description of the growth-optimal strategy $\pi^* \in \mathcal{A}$.

Theorem 4.4.2 *Suppose that Assumption* 4.3.7 *holds. Then there exists an unique growth-optimal strategy* $\pi^* \in \mathcal{A}$, *explicitly given by*

$$\pi_t^* = \left(\sigma_t\sigma_t'\right)^{-1}\sigma_t\theta_t, \tag{4.15}$$

where the process $\theta = (\theta_t)_{0 \le t \le T}$ is the market price of risk introduced in Definition 4.3.5. The corresponding Growth-Optimal Portfolio (GOP) process $V^{\pi^*} = (V_t^{\pi^*})_{0 \le t \le T}$ satisfies the following dynamics:

$$\frac{dV_t^{\pi^*}}{V_t^{\pi^*}} = r_t\,dt + \theta_t'(\theta_t\,dt + dW_t). \tag{4.16}$$

Proof Let $\pi \in \mathcal{A}$ be a trading strategy. A simple application of Itô's formula gives that

$$d\log V_t^{\pi} = g_t^{\pi}\,dt + \pi_t'\sigma_t\,dW_t, \tag{4.17}$$

where $g_t^{\pi} := r_t + \pi_t'(\mu_t - r_t\mathbf{1}) - \frac{1}{2}\pi_t'\sigma_t\sigma_t'\pi_t$ for $t \in [0,T]$. Since the matrix $\sigma_t\sigma_t'$ is P-a.s. positive definite for all $t \in [0,T]$, due to Assumption 4.2.1, a trading strategy $\pi^* \in \mathcal{A}$ is growth-optimal (in the sense of Definition 4.4.1) if and only if, for every $t \in [0,T]$, π_t^* solves the first-order condition obtained by differentiating g_t^{π} with respect to π_t. This means that π_t^* must satisfy the following condition, for every $t \in [0,T]$:

$$\mu_t - r_t\mathbf{1} - \sigma_t\sigma_t'\pi_t^* = 0.$$

Due to Assumption 4.2.1, the matrix $\sigma_t\sigma_t'$ is P-a.s. invertible for all $t \in [0,T]$. So, using Definition 4.3.5, we get the following unique optimiser π_t^*:

$$\pi_t^* = \left(\sigma_t\sigma_t'\right)^{-1}(\mu_t - r_t\mathbf{1}) = \left(\sigma_t\sigma_t'\right)^{-1}\sigma_t\theta_t \quad \text{for } t \in [0,T].$$

We now need to verify that $\pi^* = (\pi_t^*)_{0 \le t \le T} \in \mathcal{A}$. Due to Lemma 4.3.21, it suffices to check that $\int_0^T \|\sigma_t'\pi_t^*\|^2\,dt < \infty$ P-a.s. To show this, it is enough to notice that

$$\begin{aligned}
\int_0^T \|\sigma_t'\pi_t^*\|^2\,dt &= \int_0^T (\mu_t - r_t\mathbf{1})'\left(\sigma_t\sigma_t'\right)^{-1}(\mu_t - r_t\mathbf{1})\,dt \\
&= \int_0^T \|\theta_t\|^2\,dt < \infty \quad P\text{-a.s.}
\end{aligned}$$

due to Assumption 4.3.7. We have thus shown that π^* maximises the growth rate and is an admissible trading strategy. Finally, note that Eq. (4.17) leads to

$$\begin{aligned}
d\log V_t^{\pi^*} &= g_t^{\pi^*}\,dt + \left(\pi_t^*\right)'\sigma_t\,dW_t \\
&= r_t\,dt + \theta_t'\sigma_t'\left(\sigma_t\sigma_t'\right)^{-1}(\mu_t - r_t\mathbf{1})\,dt \\
&\quad - \frac{1}{2}\theta_t'\sigma_t'\left(\sigma_t\sigma_t'\right)^{-1}\sigma_t\sigma_t'\left(\sigma_t\sigma_t'\right)^{-1}\sigma_t\theta_t\,dt + \theta_t'\sigma_t'\left(\sigma_t\sigma_t'\right)^{-1}\sigma_t\,dW_t \\
&= \left(r_t + \frac{1}{2}\|\theta_t\|^2\right)dt + \theta_t'\,dW_t,
\end{aligned}$$

where the last equality is obtained by replacing θ_t with its expression as given in Definition 4.3.5. Equation (4.16) then follows by a simple application of Itô's formula. □

Remark 4.4.3

1. Results analogous to Theorem 4.4.2 can be found in Sect. 2 of Galesso and Runggaldier (2010), Example 3.7.9 of Karatzas and Shreve (1998), Sect. 2.7 of Platen (2002), Sect. 3.2 of Platen (2006), Sect. 10.2 of Platen and Heath (2006) and Proposition 2 of Platen and Runggaldier (2007). However, in all these works the growth-optimal strategy has been derived for the specific case of a complete financial market, i.e. under the additional assumptions that $d = N$ and $\mathbb{F} = \mathbb{F}^W$ (see Sect. 4.5). Here, we have instead chosen to deal with the more general situation described in Sect. 4.2, i.e. with a general incomplete market. Furthermore, we rigorously check the admissibility of the candidate growth-optimal strategy.
2. Due to Corollary 4.3.19, Assumption 4.3.7 is equivalent to the absence of arbitrages of the first kind. However, it is worth emphasising that Theorem 4.4.2 does not rely on the existence of an ELMM for the financial market $(S^0, S^1, \dots, S^N)$.
3. Due to Eq. (4.16), the discounted GOP process $\bar{V}^{\pi^*} = (\bar{V}_t^{\pi^*})_{0 \leq t \leq T}$ satisfies the following dynamics:

$$\frac{d\bar{V}_t^{\pi^*}}{\bar{V}_t^{\pi^*}} = \|\theta_t\|^2\, dt + \theta_t'\, dW_t. \tag{4.18}$$

We can immediately observe that the drift coefficient is the "square" of the diffusion coefficient, thus showing that there is a strong link between instantaneous rate of return and volatility in the GOP dynamics. Moreover, the market price of risk plays a key role in the GOP dynamics (to this effect, compare the discussion in Platen and Heath 2006, Chap. 13). Observe also that Assumption 4.3.7 is equivalent to requiring that the solution $\bar{V}^{\pi^*}$ to the SDE (4.18) is well defined and P-a.s. finite valued, meaning that the discounted GOP does not explode in the finite time interval $[0, T]$. Indeed, it can be shown, and this holds true in general semimartingale models, that the existence of a non-explosive GOP is in fact *equivalent* to the absence of arbitrages of the first kind, as can be deduced by combining Theorem 4.3.17 and Karatzas and Kardaras (2007), Theorem 4.12 (see also Christensen and Larsen 2007 and Hulley and Schweizer 2010).

Example (The classical Black–Scholes model) In order to develop an intuitive feeling for some of the concepts introduced in this section, let us briefly consider the case of the classical Black–Scholes model, i.e. a financial market represented by (S^0, S) with $r_t \equiv r$ for some $r \in \mathbb{R}$ for all $t \in [0, T]$ and $S = (S_t)_{0 \leq t \leq T}$ a real-valued process satisfying the following SDE:

$$dS_t = S_t \mu\, dt + S_t \sigma\, dW_t, \qquad S_0 = s \in (0, \infty),$$

with $\mu \in \mathbb{R}$ and $\sigma \in \mathbb{R} \setminus \{0\}$. The market price of risk process $\theta = (\theta_t)_{0 \leq t \leq T}$ is then given by $\theta_t \equiv \theta := \frac{\mu - r}{\sigma}$ for all $t \in [0, T]$. Due to Theorem 4.4.2, the GOP strategy

$\pi^* = (\pi_t^*)_{0 \le t \le T}$ is then given by $\pi_t^* \equiv \pi^* := \frac{\mu - r}{\sigma^2}$ for all $t \in [0, T]$. In this special case, Novikov's condition implies that $\widehat{Z}$ is a true martingale, yielding the density process of the (minimal) martingale measure $\widehat{Q}$ (see Remark 4.3.20).

The remaining part of this section is devoted to the derivation of some basic but fundamental properties of the GOP. Let us start with the following simple proposition.

Proposition 4.4.4 *Suppose that Assumption* 4.3.7 *holds. Then the discounted GOP process* $\bar{V}^{\pi^*} = (\bar{V}_t^{\pi^*})_{0 \le t \le T}$ *is related to the martingale deflator* $\widehat{Z} = (\widehat{Z}_t)_{0 \le t \le T}$ *as follows, for all* $t \in [0, T]$:

$$\bar{V}_t^{\pi^*} = \frac{1}{\widehat{Z}_t}.$$

Proof Assumption 4.3.7 ensures that the process $\widehat{Z} = (\widehat{Z}_t)_{0 \le t \le T}$ is P-a.s. strictly positive and well defined as a martingale deflator. Furthermore, due to Theorem 4.4.2, the growth-optimal strategy $\pi^* \in \mathcal{A}$ exists and is explicitly given by (4.15). Now it suffices to observe that, due to Eqs. (4.18) and (4.8),

$$\bar{V}_t^{\pi^*} = \exp\left(\int_0^t \theta_u' \, dW_u + \frac{1}{2} \int_0^t \|\theta_u\|^2 \, du\right) = \frac{1}{\widehat{Z}_t}. \qquad \square$$

We then immediately obtain the following corollary.

Corollary 4.4.5 *Suppose that Assumption* 4.3.7 *holds. Then, for any trading strategy* $\pi \in \mathcal{A}$, *the process* $\hat{V}^{\pi} = (\hat{V}_t^{\pi})_{0 \le t \le T}$ *defined by* $\hat{V}_t^{\pi} := V_t^{\pi} / V_t^{\pi^*}$ *for* $t \in [0, T]$ *is a non-negative local martingale and, hence, a supermartingale.*

Proof Passing to discounted quantities, we have $\hat{V}_t^{\pi} = V_t^{\pi} / V_t^{\pi^*} = \bar{V}_t^{\pi} / \bar{V}_t^{\pi^*}$. The claim then follows by combining Proposition 4.4.4 with part (b) of Proposition 4.3.9. $\square$

In order to give a better interpretation to the preceding corollary, let us give the following definition, which we adapt from Becherer (2001), Karatzas and Kardaras (2007) and Platen (2011).

Definition 4.4.6 An admissible portfolio process $V^{\tilde{\pi}} = (V_t^{\tilde{\pi}})_{0 \le t \le T}$ has the *numéraire property* if all admissible portfolio processes $V^{\pi} = (V_t^{\pi})_{0 \le t \le T}$, when denominated in units of $V^{\tilde{\pi}}$, are supermartingales, i.e. if the process $V^{\pi} / V^{\tilde{\pi}} = (V_t^{\pi} / V_t^{\tilde{\pi}})_{0 \le t \le T}$ is a supermartingale for all $\pi \in \mathcal{A}$.

The following proposition shows that if a numéraire portfolio exists, then it is also unique.

Proposition 4.4.7 *The numéraire portfolio process $V^{\tilde{\pi}} = (V_t^{\tilde{\pi}})_{0 \le t \le T}$ is unique (in the sense of indistinguishability). Furthermore, there exists an unique trading strategy $\tilde{\pi} \in \mathcal{A}$ such that $V^{\tilde{\pi}}$ is the numéraire portfolio, up to a null subset of $\Omega \times [0, T]$.*

Proof Let us first prove that if $M = (M_t)_{0 \le t \le T}$ is a P-a.s. strictly positive supermartingale such that $\frac{1}{M}$ is also a supermartingale, then $M_t = M_0$ P-a.s. for all $t \in [0, T]$. In fact, for any $0 \le s \le t \le T$,

$$
\begin{aligned}
1 = \frac{M_s}{M_s} &\ge \frac{1}{M_s} \mathbb{E}[M_t | \mathcal{F}_s] \ge \mathbb{E}\left[\frac{1}{M_t}\middle| \mathcal{F}_s\right] \mathbb{E}[M_t | \mathcal{F}_s] \\
&\ge \frac{1}{\mathbb{E}[M_t | \mathcal{F}_s]} \mathbb{E}[M_t | \mathcal{F}_s] = 1 \quad P\text{-a.s.,}
\end{aligned}
$$

where the first inequality follows from the supermartingale property of M, the second from the supermartingale property of $\frac{1}{M}$, and the third from Jensen's inequality. Hence, both M and $\frac{1}{M}$ are martingales. Furthermore, since we have $\mathbb{E}[\frac{1}{M_t} | \mathcal{F}_s] = \frac{1}{\mathbb{E}[M_t | \mathcal{F}_s]}$ and the function $x \mapsto x^{-1}$ is strictly convex on $(0, \infty)$, again Jensen's inequality implies that M_t is $\mathcal{F}_s$-measurable for all $0 \le s \le t \le T$. For $s = 0$, this implies that $M_t = \mathbb{E}[M_t | \mathcal{F}_0] = M_0$ P-a.s. for all $t \in [0, T]$.

Suppose now there exist two elements $\tilde{\pi}^1, \tilde{\pi}^2 \in \mathcal{A}$ such that both $V^{\tilde{\pi}^1}$ and $V^{\tilde{\pi}^2}$ have the numéraire property. By Definition 4.4.6, both $V^{\tilde{\pi}^1} / V^{\tilde{\pi}^2}$ and $V^{\tilde{\pi}^2} / V^{\tilde{\pi}^1}$ are P-a.s. strictly positive supermartingales. Hence, it must be $V_t^{\tilde{\pi}^1} = V_t^{\tilde{\pi}^2}$ P-a.s. for all $t \in [0, T]$, due to the general result just proved, and thus $V^{\tilde{\pi}^1}$ and $V^{\tilde{\pi}^2}$ are indistinguishable (see Karatzas and Shreve 1991, Sect. 1.1). In order to show that the two trading strategies $\tilde{\pi}^1$ and $\tilde{\pi}^2$ coincide, let us write as follows:

$$
\begin{aligned}
&\mathbb{E}\left[\int_0^T \left(\bar{V}_t^{\tilde{\pi}^1} \tilde{\pi}_t^1 - \bar{V}_t^{\tilde{\pi}^2} \tilde{\pi}_t^2\right)' \sigma_t \sigma_t' \left(\bar{V}_t^{\tilde{\pi}^1} \tilde{\pi}_t^1 - \bar{V}_t^{\tilde{\pi}^2} \tilde{\pi}_t^2\right) dt\right] \\
&\quad = \mathbb{E}\left[\left\langle \int \bar{V}^{\tilde{\pi}^1} (\tilde{\pi}^1)' \sigma \, dW - \int \bar{V}^{\tilde{\pi}^2} (\tilde{\pi}^2)' \sigma \, dW \right\rangle_T\right] = \mathbb{E}\left[\langle \bar{V}^{\tilde{\pi}^1} - \bar{V}^{\tilde{\pi}^2} \rangle_T\right] = 0,
\end{aligned}
$$

where we have used Eq. (4.5) and the fact that $\bar{V}^{\tilde{\pi}^1}$ and $\bar{V}^{\tilde{\pi}^2}$ are indistinguishable. Since, due to the standing Assumption 4.2.1, the matrix $\sigma_t \sigma_t'$ is P-a.s. positive definite for all $t \in [0, T]$ and $\bar{V}^{\tilde{\pi}^1}$ and $\bar{V}^{\tilde{\pi}^2}$ are indistinguishable, this implies that it must be $\tilde{\pi}_t := \tilde{\pi}_t^1 = \tilde{\pi}_t^2$ $P \otimes \ell$-a.e., thus showing the uniqueness of the strategy $\tilde{\pi} \in \mathcal{A}$. □

Remark 4.4.8 Note that the first part of Proposition 4.4.7 does not rely on any modelling assumption and, hence, is valid in full generality for any semimartingale model (compare also Becherer 2001, Sect. 4).

The following fundamental corollary makes precise the relation between the GOP, the numéraire portfolio and the viability of the financial market.

Corollary 4.4.9 *The financial market is viable, in the sense of Definition* 4.3.10, *if and only if the numéraire portfolio exists. Furthermore, if Assumption* 4.3.7 *holds, then the growth-optimal portfolio* V^{π^*} *coincides with the numéraire portfolio* $V^{\tilde{\pi}}$, *and the corresponding trading strategies* $\pi^*, \tilde{\pi} \in \mathcal{A}$ *coincide up to a null subset of* $\Omega \times [0,T]$.

Proof If the financial market is viable, Corollary 4.3.19 implies that Assumption 4.3.7 is satisfied. Hence, due to Theorem 4.4.2 together with Corollary 4.4.5 and Definition 4.4.6, the GOP exists and possesses the numéraire property. Conversely, suppose that the numéraire portfolio $V^{\tilde{\pi}}$ exists. Then, due to Definition 4.4.6, the process $V^{\pi}/V^{\tilde{\pi}} = (V_t^{\pi}/V_t^{\tilde{\pi}})_{0 \le t \le T}$ is a supermartingale for every $\pi \in \mathcal{A}$. In turn, this implies that $E[\bar{V}_T^{\pi}/\bar{V}_T^{\tilde{\pi}}] \le E[\bar{V}_0^{\pi}/\bar{V}_0^{\tilde{\pi}}] = 1$ for all $\pi \in \mathcal{A}$, thus showing that the set $\{\bar{V}_T^{\pi}/\bar{V}_T^{\tilde{\pi}} : \pi \in \mathcal{A}\}$ is bounded in L^1 and, hence, also in probability. Since the multiplication by the fixed random variable $\bar{V}_T^{\tilde{\pi}}$ does not affect the boundedness in probability, this implies that the NUPBR condition holds. Hence, due to Theorem 4.3.17, the financial market is viable. The second assertion follows immediately from Proposition 4.4.7. □

We emphasise again that all these results hold true even in the absence of an ELMM. For further comments on the relations between the GOP and the numéraire portfolio in a general semimartingale setting, we refer to Sect. 3 of Karatzas and Kardaras (2007) (see also Hulley and Schweizer 2010 in the continuous semimartingale case).

Remark 4.4.10 (On the GOP-denominated market) Due to Corollary 4.4.9, the GOP coincides with the numéraire portfolio. Moreover, Corollary 4.4.5 shows that all portfolio processes V^{π}, $\pi \in \mathcal{A}$, are local martingales when denominated in units of the GOP V^{π^*}. This means that, if we express all price processes in terms of the GOP, then the original probability measure P becomes an ELMM for the GOP-denominated market. Hence, due to the fundamental theorem of asset pricing (see Delbaen and Schachermayer 1994), the classical (NFLVR) no-arbitrage-type condition holds for the GOP-denominated market. This observation suggests that the GOP-denominated market may be regarded as the minimal and natural setting for dealing with valuation and portfolio optimisation problems, even when there does not exist an ELMM for the original market $(S^0, S^1, \dots, S^N)$, and this fact will be exploited in Sect. 4.6. In a related context, see also Christensen and Larsen (2007).

According to Platen (2002, 2006, 2011) and Platen and Heath (2006), let us give the following definition.

Definition 4.4.11 For any portfolio process V^{π}, the process $\hat{V}^{\pi} = (\hat{V}_t^{\pi})_{0 \le t \le T}$, defined as $\hat{V}_t^{\pi} := V_t^{\pi}/V_t^{\pi^*}$ for $t \in [0,T]$, is called the *benchmarked* portfolio process. A trading strategy $\pi \in \mathcal{A}$ and the associated portfolio process V^{π} are said to be *fair* if the benchmarked portfolio process $\hat{V}^{\pi}$ is a martingale. We denote by $\mathcal{A}^F$ the set of all fair trading strategies in $\mathcal{A}$.

According to Definition 4.4.11, the result of Corollary 4.4.5 amounts to saying that all benchmarked portfolio processes are positive supermartingales. Note that every benchmarked portfolio process is a local martingale but not necessarily a true martingale. This amounts to saying that there may exist *unfair* portfolios, namely portfolios for which the benchmarked value process is a strict local martingale. The concept of benchmarking will become relevant in Sect. 4.6.1, where we shall discuss its role for valuation purposes.

Remark 4.4.12 (Other optimality properties of the GOP) Besides maximising the growth rate, the GOP enjoys several other optimality properties, many of which are illustrated in the monograph Platen and Heath (2006). In particular, it has been shown that the GOP maximises the long-term growth rate among all admissible portfolios, see e.g. Platen (2011). It is also well known that the GOP is the solution to the problem of maximising an expected logarithmic utility function, see Sect. 4.6.3 and also Karatzas and Kardaras (2007). Other interesting properties of the GOP include the impossibility of *relative arbitrages* (or *systematic outperformance*) with respect to it, see Fernholz and Karatzas (2009) and Platen (2011), and, under suitable assumptions on the behaviour of market participants, *two-fund separation* results and connections with mean-variance efficiency, see e.g. Platen (2002, 2006). Other properties of the growth-optimal strategy are also illustrated in the recent paper MacLean et al. (2010).

4.5 Replicating Strategies and Completeness of the Financial Market

In this section we start laying the foundations for the valuation of arbitrary contingent claims without relying on the existence of an ELMM for the financial market $(S^0, S^1, \dots, S^N)$. More specifically, in this section we shall be concerned with the study of replicating (or hedging) strategies, formally defined as follows.

Definition 4.5.1 Let H be a positive $\mathcal{F}$-measurable contingent claim (i.e. random variable) such that $E[\widehat{Z}_T H/S_T^0] < \infty$. If there exists a couple $(v^H, \pi^H) \in (0, \infty) \times \mathcal{A}$ such that $V_T^{v^H, \pi^H} = H$ P-a.s., then we say that π^H is a *replicating strategy* for H.

The following proposition illustrates some basic features of a replicating strategy.

Proposition 4.5.2 *Suppose that Assumption* 4.3.7 *holds. Let* H *be a positive* $\mathcal{F}$*-measurable contingent claim such that* $\mathbb{E}[\widehat{Z}_T H/S_T^0] < \infty$ *and suppose there exists a trading strategy* $\pi^H \in \mathcal{A}$ *such that* $V_T^{v^H, \pi^H} = H$ P*-a.s. for* $v^H = \mathbb{E}[\widehat{Z}_T H/S_T^0]$. *Then the following hold*:

(a) *the strategy* π^H *is fair in the sense of Definition* 4.4.11;
(b) *the strategy* π^H *is unique up to a null subset of* $\Omega \times [0, T]$.

Moreover, for every $(v, \pi) \in (0, \infty) \times \mathcal{A}$ *such that* $V_T^{v,\pi} = H$ P*-a.s., we have* $V_t^{v,\pi} \geq V_t^{v^H,\pi^H}$ P*-a.s. for all* $t \in [0, T]$*. In particular, there cannot exist an element* $\bar{\pi} \in \mathcal{A}$ *such that* $V_T^{\bar{v},\bar{\pi}} = H$ P*-a.s. for some* $\bar{v} < v^H$.

Proof Corollary 4.4.5 implies that the process $\hat{V}^{v^H,\pi^H} = (V_t^{v^H,\pi^H}/V_t^{\pi^*})_{0\leq t\leq T}$ is a supermartingale. Moreover, it is also a martingale, due to the fact that

$$\hat{V}_0^{v^H,\pi^H} = v^H = E\left[\frac{\widehat{Z}_T}{S_T^0}H\right] = E\left[\frac{V_T^{v^H,\pi^H}}{V_T^{\pi^*}}\right] = E\left[\hat{V}_T^{v^H,\pi^H}\right], \tag{4.19}$$

where the third equality follows from Proposition 4.4.4. Part (a) then follows from Definition 4.4.11. To prove part (b), let $\hat{\pi} \in \mathcal{A}$ be a trading strategy such that $V_T^{v^H,\hat{\pi}} = H$ P-a.s. for $v^H = \mathbb{E}[\widehat{Z}_T H/S_T^0]$. Reasoning as in (4.19), the benchmarked portfolio process $\hat{V}^{v^H,\hat{\pi}} = (V_t^{v^H,\hat{\pi}}/V_t^{\pi^*})_{0\leq t\leq T}$ is a martingale. Together with the fact that $\hat{V}_T^{v^H,\hat{\pi}} = \widehat{Z}_T H/S_T^0 = \hat{V}_T^{v^H,\pi^H}$ P-a.s., this implies that $V_t^{v^H,\hat{\pi}} = V_t^{v^H,\pi^H}$ P-a.s. for all $t \in [0, T]$. Part (b) then follows by the same arguments as in the second part of the proof of Proposition 4.4.7. To prove the last assertion, let $(v, \pi) \in (0, \infty) \times \mathcal{A}$ be such that $V_T^{v,\pi} = H$ P-a.s. Due to Corollary 4.4.5, the benchmarked portfolio process $\hat{V}^{v,\pi} = (V_t^{v,\pi}/V_t^{\pi^*})_{0\leq t\leq T}$ is a supermartingale. So, for any $t \in [0, T]$, due to part (a),

$$\hat{V}_t^{v^H,\pi^H} = E\left[\hat{V}_T^{v^H,\pi^H}\middle|\mathcal{F}_t\right] = E\left[\frac{\widehat{Z}_T}{S_T^0}H\middle|\mathcal{F}_t\right] = E\left[\hat{V}_T^{v,\pi}\middle|\mathcal{F}_t\right] \leq \hat{V}_t^{v,\pi} \quad P\text{-a.s.},$$

and, hence, $V_t^{v^H,\pi^H} \leq V_t^{v,\pi}$ P-a.s. for all $t \in [0, T]$. For $t = 0$, this implies that $v \geq v^H$, thus completing the proof. □

Remark 4.5.3 Observe that Proposition 4.5.2 does not exclude the existence of a trading strategy $\check{\pi} \in \mathcal{A}$ such that $V_T^{\check{v},\check{\pi}} = H$ P-a.s. for some $\check{v} > v^H$. However, one can argue that it may not be optimal to invest in such a strategy in order to replicate H, since it requires a larger initial investment and leads to an unfair portfolio process. Indeed, Proposition 4.5.2 shows that $v^H = E[\widehat{Z}_T H/S_T^0]$ is the minimal initial capital starting from which one can replicate the contingent claim H. To this effect, see also Remark 1.6.4 in Karatzas and Shreve (1998).

A particularly nice and interesting situation arises when the financial market is *complete*, meaning that every *contingent claim* can be perfectly replicated starting from some initial investment by investing in the financial market according to some admissible self-financing trading strategy.

Definition 4.5.4 The financial market $(S^0, S^1, \dots, S^N)$ is said to be *complete* if for any positive $\mathcal{F}$-measurable contingent claim H such that $\mathbb{E}[\widehat{Z}_T H/S^0_T] < \infty$, there exists a couple $(v^H, \pi^H) \in (0,\infty) \times \mathcal{A}$ such that $V_T^{v^H,\pi^H} = H$ P-a.s.

In general, the financial market described in Sect. 4.2 is incomplete, and, hence, not all contingent claims can be perfectly replicated. The following theorem gives a sufficient condition for the financial market to be complete. The proof is similar to that of Theorem 1.6.6 in Karatzas and Shreve (1998), except that we avoid the use of any ELMM, since the latter may fail to exist in our general context. This allows us to highlight the fact that the concept of market completeness does not depend on the existence of an ELMM.

Theorem 4.5.5 *Suppose that Assumption* 4.3.7 *holds. If* $\mathbb{F} = \mathbb{F}^W$, *where* $\mathbb{F}^W$ *denotes the* P*-augmented Brownian filtration associated to* W, *and* $d = N$, *then the financial market* $(S^0, S^1, \dots, S^N)$ *is complete. More precisely, any positive* $\mathcal{F}$*-measurable contingent claim* H *with* $E[\widehat{Z}_T H/S^0_T] < \infty$ *can be replicated by a fair portfolio process* V^{v^H,π^H} *with* $v^H = E[\widehat{Z}_T H/S^0_T]$ *and* $\pi^H \in \mathcal{A}^F$.

Proof Let H be a positive $\mathcal{F} = \mathcal{F}^W_T$-measurable random variable such that $\mathbb{E}[\widehat{Z}_T H/S^0_T] < \infty$ and define the martingale $M = (M_t)_{0 \le t \le T}$ by $M_t := \mathbb{E}[\widehat{Z}_T H/S^0_T | \mathcal{F}_t]$ for $t \in [0,T]$. According to the martingale representation theorem (see Karatzas and Shreve 1991, Theorem 3.4.15 and Problem 3.4.16), there exists an $\mathbb{R}^N$-valued progressively measurable process $\varphi = (\varphi_t)_{0 \le t \le T}$ such that $\int_0^T \|\varphi_t\|^2\, dt < \infty$ P-a.s. and

$$M_t = M_0 + \int_0^t \varphi'_u\, dW_u \quad \text{for all } t \in [0,T]. \tag{4.20}$$

Define then the positive process $V = (V_t)_{0 \le t \le T}$ by $V_t := \frac{S^0_t}{\widehat{Z}_t} M_t$ for $t \in [0,T]$. Recalling that $S^0_0 = 1$, we have $v^H := V_0 = M_0 = \mathbb{E}[\widehat{Z}_T H/S^0_T]$. The standing Assumption 4.2.1, together with the fact that $d = N$, implies that the matrix σ_t is P-a.s. invertible for all $t \in [0,T]$. Then, an application of the product rule together with Eqs. (4.8) and (4.20), gives

$$
\begin{aligned}
d\left(\frac{V_t}{S_t^0}\right) &= d\left(\frac{M_t}{\widehat{Z}_t}\right) = M_t\, d\frac{1}{\widehat{Z}_t} + \frac{1}{\widehat{Z}_t}\, dM_t + d\left\langle M, \frac{1}{\widehat{Z}}\right\rangle_t \\
&= \frac{M_t}{\widehat{Z}_t}\theta_t'\, dW_t + \frac{M_t}{\widehat{Z}_t}\,\|\theta_t\|^2\, dt + \frac{1}{\widehat{Z}_t}\varphi_t'\, dW_t + \frac{1}{\widehat{Z}_t}\varphi_t'\theta_t\, dt \\
&= \frac{V_t}{S_t^0}\left(\theta_t + \frac{\varphi_t}{M_t}\right)'\theta_t\, dt + \frac{V_t}{S_t^0}\left(\theta_t + \frac{\varphi_t}{M_t}\right)' dW_t \\
&= \frac{V_t}{S_t^0}\left(\theta_t + \frac{\varphi_t}{M_t}\right)'\sigma_t^{-1}(\mu_t - r_t\mathbf{1})\, dt + \frac{V_t}{S_t^0}\left(\theta_t + \frac{\varphi_t}{M_t}\right)'\sigma_t^{-1}\,\sigma_t\, dW_t \\
&= \frac{V_t}{S_t^0}\sum_{i=1}^N \pi_t^{H,i}\frac{d\bar{S}_t^i}{\bar{S}_t^i},
\end{aligned}
\tag{4.21}
$$

where $\pi_t^H = (\pi_t^{H,1}, \ldots, \pi_t^{H,N})' := (\sigma_t')^{-1}(\theta_t + \frac{\varphi_t}{M_t})$ for all $t \in [0, T]$. The last line of (4.21) shows that the process $\bar{V} := V/S^0 = (V_t/S_t^0)_{0\le t\le T}$ can be represented as a stochastic exponential as in part (b) of Definition 4.2.3. Hence, it remains to check that the process π^H satisfies the integrability conditions of part (a) of Definition 4.2.3. Due to Lemma 4.3.21, it suffices to verify that $\int_0^T \|\sigma_t'\pi_t^H\|^2\, dt < \infty$ P-a.s. This can be shown as follows:

$$
\begin{aligned}
\int_0^T \|\sigma_t'\pi_t^H\|^2\, dt &= \int_0^T \left\|\theta_t + \frac{\varphi_t}{M_t}\right\|^2 dt \\
&\le 2\int_0^T \|\theta_t\|^2\, dt + 2\left\|\frac{1}{M}\right\|_\infty \int_0^T \|\varphi_t\|^2\, dt < \infty \quad P\text{-a.s.}
\end{aligned}
$$

due to Assumption 4.3.7 and because $\|\frac{1}{M}\|_\infty := \max_{t\in[0,T]} |\frac{1}{M_t}| < \infty$ P-a.s. by the continuity of M. We have thus shown that π^H is an admissible trading strategy, i.e. $\pi^H \in \mathcal{A}$, and the associated portfolio process $V^{v^H,\pi^H} = (V_t^{v^H,\pi^H})_{0\le t\le T}$ satisfies $V_T^{v^H,\pi^H} = V_T = H$ P-a.s. with $v^H = \mathbb{E}[\widehat{Z}_T H/S_T^0]$. The fact that $\pi^H \in \mathcal{A}^F$ follows from the equality $\hat{V}_t^{v^H,\pi^H} = V_t^{v^H,\pi^H}/V_t^{\pi^*} = V_t\,\widehat{Z}_t/S_t^0 = M_t$. □

We close this section with some important comments on the result of Theorem 4.5.5.

Remark 4.5.6

1. We want to emphasise that Theorem 4.5.5 does not rely on the existence of an ELMM for the financial market $(S^0, S^1, \ldots, S^N)$. This amounts to saying that the completeness of a financial market does not necessarily imply that some mild forms of arbitrage opportunities are a priori excluded. Typical "textbook versions" of the so-called *second Fundamental Theorem of Asset Pricing* state that the completeness of the financial market is equivalent to the uniqueness of

the *Equivalent (Local) Martingale Measure*, loosely speaking. However, Theorem 4.5.5 shows that we can have a complete financial market even when no E(L)MM exists at all. The fact that absence of arbitrage opportunities and market completeness should be regarded as distinct concepts has been already pointed out in a very general setting in Jarrow and Madan (1999). The completeness of the financial market model will play a crucial role in Sect. 4.6, where we shall be concerned with valuation and hedging problems in the absence of an ELMM.

2. Following the reasoning in the proof of Theorem 1.6.6 of Karatzas and Shreve (1998), but avoiding the use of an ELMM (which in our context may fail to exist), it is possible to prove a converse result to Theorem 4.5.5. More precisely, if we assume that $\mathbb{F} = \mathbb{F}^W$ and that every $\mathcal{F}$-measurable positive random variable H with $v^H := E[\widehat{Z}_T H / S_T^0] < \infty$ admits a trading strategy $\pi^H \in \mathcal{A}$ such that $V_T^{v^H, \pi^H} = H$ P-a.s., then we necessarily have $d = N$. Moreover, it can be shown that the completeness of the financial market is *equivalent* to the existence of a unique martingale deflator, and this holds true even in more general models based on continuous semimartingales. For details, we refer the interested reader to Chap. 4 of Fontana (2012).

4.6 Contingent Claim Valuation Without ELMMs

The main goal of this section is to show how one can proceed to the valuation of contingent claims in financial market models which may not necessarily admit an ELMM. Since the non-existence of a properly defined martingale measure precludes the whole machinery of risk-neutral pricing, this appears as a non-trivial issue. Here we concentrate on the situation of a complete financial market, as considered at the end of the last section (see Sect. 4.7 for possible extensions to incomplete markets). A major focus of this section is on providing a mathematical justification for the so-called *real-world pricing approach*, according to which the valuation of contingent claims is performed under the original (or *real-world*) probability measure P using the GOP as the natural numéraire.

Remark 4.6.1 In this section we shall be concerned with the problem of *pricing* contingent claims. However, one should be rather careful with the terminology and distinguish between a *value* assigned to a contingent claim and its prevailing *market price*. Indeed, the former represents the outcome of an a priori chosen valuation rule, while the latter is the price determined by supply and demand forces in the financial market. Since the choice of the valuation criterion is a subjective one, the two concepts of *value* and *market price* do not necessarily coincide. This is especially true when arbitrage opportunities and/or bubble phenomena are not excluded from the financial market. In this section, we use the word "price" only to be consistent with the standard terminology in the literature.

4.6.1 Real-World Pricing and the Benchmark Approach

We start by introducing the concept of *real-world price*, which is at the core of the so-called *benchmark approach* to the valuation of contingent claims.

Definition 4.6.2 Let H be a positive $\mathcal{F}$-measurable contingent claim such that $\mathbb{E}[\widehat{Z}_T H/S_T^0] < \infty$. If there exists a fair portfolio process $V^{v^H,\pi^H} = (V_t^{v^H,\pi^H})_{0\le t\le T}$ such that $V_T^{v^H,\pi^H} = H$ P-a.s. for some $(v^H,\pi^H) \in (0,\infty)\times\mathcal{A}^F$, then the *real-world price* of H at time t, denoted as Π_t^H, is defined as follows:

$$\Pi_t^H := V_t^{\pi^*}\mathbb{E}\left[\frac{H}{V_T^{\pi^*}}\middle|\mathcal{F}_t\right] \tag{4.22}$$

for every $t \in [0,T]$, where $V^{\pi^*} = (V_t^{\pi^*})_{0\le t\le T}$ denotes the GOP.

The terminology *real-world price* is used to indicate that, unlike in the traditional setting, all contingent claims are valued under the original real-world probability measure P and not under an equivalent risk-neutral measure. This allows us to extend the valuation of contingent claims to financial markets for which no ELMM may exist. The concept of *real-world price* gives rise to the so-called *benchmark approach* to the valuation of contingent claims in view of the fact that the GOP plays the role of the natural numéraire portfolio (compare Remark 4.4.10). For this reason, we shall refer to it as the *benchmark* portfolio. We refer the reader to Platen (2006, 2011) and Platen and Heath (2006) for a thorough presentation of the benchmark approach.

Clearly, if there exists a fair portfolio process V^{v^H,π^H} such that $V_T^{v^H,\pi^H} = H$ P-a.s. for $(v^H,\pi^H) \in (0,\infty)\times\mathcal{A}^F$, then the real-world price coincides with the value of the fair replicating portfolio. In fact, for all $t \in [0,T]$,

$$\Pi_t^H = V_t^{\pi^*} E\left[\frac{H}{V_T^{\pi^*}}\middle|\mathcal{F}_t\right] = V_t^{\pi^*} E\left[\frac{V_T^{v^H,\pi^H}}{V_T^{\pi^*}}|\mathcal{F}_t\right] = V_t^{v^H,\pi^H} \quad P\text{-a.s.},$$

where the last equality is due to the fairness of V^{v^H,π^H}, see Definition 4.4.11. Moreover, the second part of Proposition 4.5.2 gives an economic rationale for the use of the real-world pricing formula (4.22), since it shows that the latter gives the value of the least expensive replicating portfolio. This property has been called the *law of the minimal price* (see Platen 2011, Sect. 4). The following simple proposition immediately follows from Theorem 4.5.5.

Proposition 4.6.3 *Suppose that Assumption* 4.3.7 *holds. Let H be a positive $\mathcal{F}$-measurable contingent claim such that $\mathbb{E}[\widehat{Z}_T H/S_T^0] < \infty$. Then, under the assumptions of Theorem* 4.5.5, *the following hold*:

(a) *there exists a fair portfolio process* $V^{v^H,\pi^H} = (V_t^{v^H,\pi^H})_{0\le t\le T}$ *such that* $V_T^{v^H,\pi^H} = H$ *P-a.s.*;
(b) *the real-world price of* H *(at time* $t=0$*) is given by*

$$\Pi_0^H = E\left[\frac{H}{V_T^{\pi*}}\right] = E\left[\frac{\widehat{Z}_T}{S_T^0}H\right] = v^H.$$

Remark 4.6.4

1. Notice that, due to Proposition 4.4.4, the real-world pricing formula (4.22) can be rewritten as follows, for any $t \in [0,T]$:

$$\Pi_t^H = \frac{S_t^0}{\widehat{Z}_t}\mathbb{E}\left[\frac{\widehat{Z}_T}{S_T^0}H\middle|\mathcal{F}_t\right]. \tag{4.23}$$

Suppose now that $\mathbb{E}[\widehat{Z}_T] = 1$, so that $\widehat{Z}$ represents the density process of the ELMM $\widehat{Q}$ (see Remark 4.3.20). Due to the Bayes formula, Eq. (4.23) can then be rewritten as follows:

$$\Pi_t^H = S_t^0\mathbb{E}^{\widehat{Q}}\left[\frac{H}{S_T^0}\middle|\mathcal{F}_t\right],$$

and we recover the usual risk-neutral pricing formula (see also Platen 2011, Sect. 5, and Platen and Heath 2006, Sect. 10.4). In this sense, the real-world pricing approach can be regarded as a consistent extension of the usual risk-neutral valuation approach to a financial market for which an ELMM may fail to exist.
2. Let us suppose for a moment that H and the final value of the GOP $V_T^{\pi^*}$ are conditionally independent given the σ-field $\mathcal{F}_t$, for all $t \in [0,T]$. The real-world pricing formula (4.22) can then be rewritten as follows:

$$\Pi_t^H = V_t^{\pi^*}\,\mathbb{E}\left[\frac{1}{V_T^{\pi*}}\middle|\mathcal{F}_t\right]\mathbb{E}[H|\mathcal{F}_t] =: P(t,T)\mathbb{E}[H|\mathcal{F}_t], \tag{4.24}$$

where $P(t,T)$ denotes the *fair value* at time t of a *zero coupon bond* with maturity T (i.e. a contingent claim which pays the deterministic amount 1 at time T). This shows that, under the (rather strong) assumption of conditional independence, one can recover the well-known *actuarial pricing formula* (see also Platen 2006, Corollary 3.4, and Platen 2011, Sect. 5).
3. We want to point out that part (b) of Proposition 4.6.3 can be easily generalised to any time $t \in [0,T]$; compare for instance Proposition 10 in Galesso and Runggaldier (2010).

In view of the above remarks, it is interesting to observe how several different valuation approaches which have been widely used in finance and insurance, such as risk-neutral pricing and actuarial pricing, are both generalised and unified under the concept of real-world pricing. We refer to Sect. 10.4 of Platen and Heath (2006) for related comments on the unifying aspects of the benchmark approach.

4.6.2 The **Upper Hedging Price** *Approach*

The *upper hedging price* (or *super-hedging price*) is a classical approach to the valuation of contingent claims (see e.g. Karatzas and Shreve 1998, Sect. 5.5.3). The intuitive idea is to find the smallest initial capital which allows one to obtain a final wealth that is greater or equal than the payoff at maturity of a given contingent claim.

Definition 4.6.5 Let H be a positive $\mathcal{F}$-measurable contingent claim. The *upper hedging price* $\mathcal{U}(H)$ of H is defined as follows:

$$\mathcal{U}(H) := \inf\bigl\{v \in [0,\infty) : \exists \pi \in \mathcal{A} \text{ such that } V_T^{v,\pi} \geq H \ P\text{-a.s.}\bigr\}$$

with the usual convention $\inf \emptyset = \infty$.

The next proposition shows that, in a complete diffusion-based financial market, the upper hedging price takes a particularly simple and natural form. This result is an immediate consequence of the supermartingale property of benchmarked portfolio processes together with the completeness of the financial market, but, for the reader's convenience, we give a detailed proof.

Proposition 4.6.6 *Let H be a positive $\mathcal{F}$-measurable contingent claim such that $\mathbb{E}[\widehat{Z}_T H/S_T^0] < \infty$. Then, under the assumptions of Theorem* 4.5.5, *the upper hedging price of H is explicitly given by*

$$\mathcal{U}(H) = \mathbb{E}\left[\frac{\widehat{Z}_T}{S_T^0} H\right]. \tag{4.25}$$

Proof In order to prove (4.25), we show both directions of inequality.

($\geq$) If $\{v \in [0,\infty) : \exists \pi \in \mathcal{A} \text{ such that } V_T^{v,\pi} \geq H \ P\text{-a.s.}\} = \emptyset$, then we have $\mathbb{E}[\widehat{Z}_T H/S_T^0] < \mathcal{U}(H) = \infty$. So, let us assume that there exists $(v,\pi) \in [0,\infty) \times \mathcal{A}$ such that $V_T^{v,\pi} \geq H$ P-a.s. Under Assumption 4.3.7, due to Corollary 4.4.5, the benchmarked portfolio process $\hat{V}^{v,\pi} = (V_t^{v,\pi}/V_t^{\pi^*})_{0 \leq t \leq T}$ is a supermartingale, and so, recalling also Proposition 4.4.4, we have

$$v = \hat{V}_0^{v,\pi} \geq \mathbb{E}\bigl[\hat{V}_T^{v,\pi}\bigr] = \mathbb{E}\left[\frac{\widehat{Z}_T}{S_T^0} V_T^{v,\pi}\right] \geq \mathbb{E}\left[\frac{\widehat{Z}_T}{S_T^0} H\right].$$

This implies that $\mathcal{U}(H) \geq \mathbb{E}[\frac{\widehat{Z}_T}{S_T^0} H]$.

($\leq$) Under the present assumptions, Theorem 4.5.5 yields the existence of a couple $(v^H, \pi^H) \in (0,\infty) \times \mathcal{A}^F$ such that $V_T^{v^H,\pi^H} = H$ P-a.s. and $v^H = \mathbb{E}[\widehat{Z}_T H/S_T^0]$. Hence,

$$\mathbb{E}\left[\frac{\widehat{Z}_T}{S_T^0} H\right] = v^H \in \bigl\{v \in [0,\infty) : \exists \pi \in \mathcal{A} \text{ such that } V_T^{v,\pi} \geq H \ P\text{-a.s.}\bigr\}.$$

This implies that $\mathcal{U}(H) \leq \mathbb{E}[\widehat{Z}_T H/S_T^0]$. □

An analogous result can be found in Proposition 5.3.2 of Karatzas and Shreve (1998) (compare also Fernholz and Karatzas 2009, Sect. 10). We want to point out that Definition 4.6.5 can be easily generalised to an arbitrary time point $t \in [0, T]$ in order to define the upper hedging price at $t \in [0, T]$. The result of Proposition 4.6.6 carries over to this slightly generalised setting with essentially the same proof, see e.g. Theorem 3 in Galesso and Runggaldier (2010).

Remark 4.6.7

1. Notice that, due to Proposition 4.4.4, Eq. (4.25) can be rewritten as follows:

$$\mathcal{U}(H) = \mathbb{E}\left[\frac{\widehat{Z}_T}{S_T^0} H\right] = \mathbb{E}\left[\frac{H}{V_T^{\pi^*}}\right].$$

   This shows that the upper hedging price can be obtained by computing the expectation of the benchmarked value (in the sense of Definition 4.4.11) of the contingent claim H under the real-world probability measure P and thus coincides with the real-world price (evaluated at $t = 0$), see part *(b)* of Proposition 4.6.3.
2. Suppose that $\mathbb{E}[\widehat{Z}_T] = 1$. As explained in Remark 4.3.20, the process $\widehat{Z}$ represents then the density process of the ELMM $\widehat{Q}$. In this case, the upper hedging price $\mathcal{U}(H)$ yields the usual risk-neutral valuation formula, i.e. we have $\mathcal{U}(H) = \mathbb{E}^{\widehat{Q}}[H/S_T^0]$.

4.6.3 Utility Indifference Valuation

The real-world valuation approach has been justified so far on the basis of replication arguments, as can be seen from Propositions 4.6.3 and 4.6.6. We now present a different approach that uses the idea of *utility indifference valuation*. To this effect, let us first consider the problem of maximising an expected utility function of the discounted final wealth. Recall that, due to Theorem 4.3.17, we can meaningfully consider portfolio optimisation problems even in the absence of an ELMM for $(S^0, S^1, \dots, S^N)$.

Definition 4.6.8 We call a function $U : [0, \infty) \to [0, \infty)$ a *utility function* if:

1. U is strictly increasing and strictly concave, continuously differentiable;
2. $\lim_{x\to\infty} U'(x) = 0$ and $\lim_{x\to 0} U'(x) = \infty$.

Problem (expected utility maximisation) Let U be as in Definition 4.6.8, and let $v \in (0, \infty)$. The expected utility maximisation problem consists in the following:

$$\text{maximise} \quad \mathbb{E}\big[U\big(\bar{V}_T^{v,\pi}\big)\big] \quad \text{over all } \pi \in \mathcal{A}. \tag{4.26}$$

The following lemma shows that, in the case of a complete financial market, there is no loss of generality in restricting our attention to fair strategies only. Recall that, due to Definition 4.4.11, $\mathcal{A}^F$ denotes the set of all fair trading strategies in $\mathcal{A}$.

Lemma 4.6.9 *Under the assumptions of Theorem* 4.5.5, *for any utility function U and for any* $v \in (0, \infty)$, *the following holds*:

$$\sup_{\pi \in \mathcal{A}} E\big[U\big(\bar{V}_T^{v,\pi}\big)\big] = \sup_{\pi \in \mathcal{A}^F} E\big[U\big(\bar{V}_T^{v,\pi}\big)\big]. \tag{4.27}$$

Proof It is clear that "$\geq$" holds in (4.27) since $\mathcal{A}^F \subseteq \mathcal{A}$. To show the reverse inequality, let us consider an arbitrary strategy $\pi \in \mathcal{A}$. The benchmarked portfolio process $\hat{V}^{v,\pi} = (V_t^{v,\pi} / V_t^{\pi^*})_{0 \leq t \leq T}$ is a supermartingale due to Corollary 4.4.5, and hence:

$$v' := E\left[\frac{\widehat{Z}_T}{S_T^0} V_T^{v,\pi}\right] = E\left[\frac{V_T^{v,\pi}}{V_T^{\pi^*}}\right] \leq v$$

with equality holding if and only if $\pi \in \mathcal{A}^F$. Let $\bar{v} := v - v' \geq 0$. It is clear that the positive $\mathcal{F}$-measurable random variable $\bar{H} := \bar{V}_T^{v,\pi} + \bar{v}/\widehat{Z}_T$ satisfies $E[\widehat{Z}_T \bar{H}] = v$, and so, due to Theorem 4.5.5, there exists an admissible trading strategy $\pi^H \in \mathcal{A}^F$ such that $\bar{V}_T^{v,\pi^H} = \bar{H} \geq \bar{V}_T^{v,\pi}$ P-a.s., with equality holding if and only if the strategy π is fair. We then have, due to the monotonicity of U,

$$E\big[U\big(\bar{V}_T^{v,\pi}\big)\big] \leq E\big[U(\bar{H})\big] = E\big[U\big(\bar{V}_T^{v,\pi^H}\big)\big] \leq \sup_{\pi \in \mathcal{A}^F} E\big[U\big(\bar{V}_T^{v,\pi}\big)\big].$$

Since $\pi \in \mathcal{A}$ was arbitrary, this shows the "$\leq$" inequality in (4.27). □

In particular, Lemma 4.6.9 shows that, in the context of portfolio optimisation problems, restricting the class of admissible trading strategies to *fair* admissible strategies is not only "reasonable", as argued in Chap. 11 of Platen and Heath (2006), but exactly yields the same optimal value of the problem in its original formulation. The following theorem gives the solution to problem (4.26) in the case of a complete financial market. Related results can be found in Lemma 5 of Galesso and Runggaldier (2010) and Theorem 3.7.6 of Karatzas and Shreve (1998).

Theorem 4.6.10 *Let the assumptions of Theorem* 4.5.5 *hold, and let U be a utility function. For* $v \in (0, \infty)$, *assume that the function* $\mathcal{W}(y) := \mathbb{E}[\widehat{Z}_T\, I(y/\bar{V}_T^{v,\pi^*})]$ *is finite for every* $y \in (0, \infty)$, *where I is the inverse function of* U'. *Then the function* $\mathcal{W}$ *is invertible, and the optimal discounted final wealth* $\bar{V}_T^{v,\pi^U}$ *for problem* (4.26) *is explicitly given as follows*:

$$\bar{V}_T^{v,\pi^U} = I\left(\frac{\mathcal{Y}(v)}{\bar{V}_T^{v,\pi^*}}\right), \tag{4.28}$$

where $\mathcal{Y}$ *denotes the inverse function of* $\mathcal{W}$. *The optimal strategy* $\pi^U \in \mathcal{A}^F$ *is given by the replicating strategy for the right-hand side of* (4.28).

Proof Note first that, due to Definition 4.6.8, the function U' admits a strictly decreasing continuous inverse function $I : [0,\infty] \to [0,\infty]$ with $I(0) = \infty$ and $I(\infty) = 0$. We have then the following well-known result from convex analysis (see e.g. Karatzas and Shreve 1998, Sect. 3.4):

$$U\big(I(y)\big) - yI(y) \geq U(x) - xy \quad \text{for } 0 \leq x < \infty,\ 0 < y < \infty. \tag{4.29}$$

As in Lemma 3.6.2 of Karatzas and Shreve (1998), it can be shown that the function $\mathcal{W} : [0,\infty] \to [0,\infty]$ is strictly decreasing and continuous, and, hence, it admits an inverse function $\mathcal{Y} : [0,\infty] \to [0,\infty]$. Since $\mathcal{W}(\mathcal{Y}(v)) = v$ for any $v \in (0,\infty)$, Theorem 4.5.5 shows that there exists a fair strategy $\pi^U \in \mathcal{A}^F$ which satisfies $\bar{V}_T^{v,\pi^U} = I(\mathcal{Y}(v)/\bar{V}_T^{v,\pi^*})$ P-a.s. Furthermore, for any $\pi \in \mathcal{A}^F$, inequality (4.29) with $y = \mathcal{Y}(v)/\bar{V}_T^{v,\pi^*}$ and $x = \bar{V}_T^{v,\pi}$ gives that

$$\begin{aligned}
E\big[U\big(\bar{V}_T^{v,\pi^U}\big)\big] &= E\bigg[U\bigg(I\bigg(\frac{\mathcal{Y}(v)}{\bar{V}_T^{v,\pi^*}}\bigg)\bigg)\bigg] \\
&\geq E\big[U\big(\bar{V}_T^{v,\pi}\big)\big] + \mathcal{Y}(v)E\bigg[\frac{1}{\bar{V}_T^{v,\pi^*}}\bigg(I\bigg(\frac{\mathcal{Y}(v)}{\bar{V}_T^{v,\pi^*}}\bigg) - \bar{V}_T^{v,\pi}\bigg)\bigg] \\
&= E\big[U\big(\bar{V}_T^{v,\pi}\big)\big] + \mathcal{Y}(v)E\bigg[\frac{1}{\bar{V}_T^{v,\pi^*}}\big(\bar{V}_T^{v,\pi^U} - \bar{V}_T^{v,\pi}\big)\bigg] \\
&= E\big[U\big(\bar{V}_T^{v,\pi}\big)\big],
\end{aligned}$$

thus showing that, based also on Lemma 4.6.9, $\pi^U \in \mathcal{A}^F$ solves problem (4.26). □

Remark 4.6.11

1. It is important to observe that Theorem 4.6.10 does not rely on the existence of an ELMM. This amounts to saying that we can meaningfully solve expected utility maximisation problems even when no ELMM exists or, equivalently, when the traditional (NFLVR) no-arbitrage-type condition fails to hold. The crucial assumption for the validity of Theorem 4.6.10 is Assumption 4.3.7, which ensures that the financial market is viable, in the sense that there are no arbitrages of the first kind (compare Theorem 4.3.17 and Corollary 4.3.19).
2. The assumption that the function $\mathcal{W}(y) := \mathbb{E}[\widehat{Z}_T\, I(y/\bar{V}_T^{v,\pi^*})]$ is finite for every $y \in (0,\infty)$ can be replaced by suitable technical conditions on the utility function U and on the processes μ and σ (see Remarks 3.6.8 and 3.6.9 in Karatzas and Shreve 1998 for more details).

Having solved in general the expected utility maximisation problem, we are now in a position to give the definition of the *utility indifference price*, in the spirit of Davis (1997) (compare also Galesso and Runggaldier 2010, Sect. 4.2; Platen and

Heath 2006, Definition 11.4.1, and Platen and Runggaldier 2007, Definition 10).[6] Until the end of this section, we let U be a utility function, in the sense of Definition 4.6.8, such that all expected values below exist and are finite.

Definition 4.6.12 Let H be a positive $\mathcal{F}$-measurable contingent claim, and let $v \in (0, \infty)$. For $p \geq 0$, let us define, for a given utility function U, the function $W_p^U : [0, 1] \to [0, \infty)$ as follows:

$$W_p^U(\varepsilon) := \mathbb{E}\big[U\big((v - \varepsilon p)\bar{V}_T^{\pi^U} + \varepsilon \bar{H}\big)\big], \tag{4.30}$$

where $\pi^U \in \mathcal{A}^F$ solves problem (4.26) for the utility function U. The *utility indifference price* of the contingent claim H is defined as the value $p(H)$ that satisfies the following condition:

$$\lim_{\varepsilon \to 0} \frac{W_{p(H)}^U(\varepsilon) - W_{p(H)}^U(0)}{\varepsilon} = 0. \tag{4.31}$$

Definition 4.6.12 is based on a "marginal rate of substitution" argument, as first pointed out in Davis (1997). In fact, $p(H)$ can be thought of as the value at which an investor is marginally indifferent between the two following alternatives:

- invest an infinitesimal part $\varepsilon p(H)$ of the initial endowment v into the contingent claim H and invest the remaining wealth $(v - \varepsilon p(H))$ according to the optimal trading strategy π^U;
- ignore the contingent claim H and simply invest the whole initial endowment v according to the optimal trading strategy π^U.

The following simple result, essentially due to Davis (1997) (compare also Platen and Heath 2006, Sect. 11.4), gives a general representation of the utility indifference price $p(H)$.

Proposition 4.6.13 *Let U be a utility function, and H a positive $\mathcal{F}$-measurable contingent claim. The utility indifference price $p(H)$ can be represented as follows*:

$$p(H) = \frac{\mathbb{E}[U'(\bar{V}_T^{v,\pi^U})\,\bar{H}]}{\mathbb{E}[U'(\bar{V}_T^{v,\pi^U})\,\bar{V}_T^{\pi^U}]}. \tag{4.32}$$

Proof Using Eq. (4.30), let us write the following Taylor's expansion:

[6]In Galesso and Runggaldier (2010) and Platen and Runggaldier (2007) the authors generalise Definition 4.6.12 to an arbitrary time $t \in [0, T]$. However, since the results and the techniques remain essentially unchanged, we only consider the basic case $t = 0$.

$$
\begin{aligned}
W_p^U(\varepsilon) &= \mathbb{E}\big[U\big(\bar{V}_T^{v,\pi^U}\big) + \varepsilon\, U'\big(\bar{V}_T^{v,\pi^U}\big)\big(\bar{H} - p\,\bar{V}_T^{\pi^U}\big) + o(\varepsilon)\big] \\
&= W_p^U(0) + \varepsilon\,\mathbb{E}\big[U'\big(\bar{V}_T^{v,\pi^U}\big)\big(\bar{H} - p\,\bar{V}_T^{\pi^U}\big)\big] + o(\varepsilon).
\end{aligned} \tag{4.33}
$$

Inserting (4.33) into (4.31), we get:

$$
\mathbb{E}\big[U'\big(\bar{V}_T^{v,\pi^U}\big)\big(\bar{H} - p(H)\,\bar{V}_T^{\pi^U}\big)\big] = 0,
$$

from which (4.32) immediately follows. □

By combining Theorem 4.6.10 with Proposition 4.6.13, we can easily prove the following corollary, which yields an explicit and "universal" representation of the utility indifference price $p(H)$ (compare also Galesso and Runggaldier 2010, Theorem 8; Platen and Heath 2006, Sect. 11.4, and Platen and Runggaldier 2007, Proposition 11).

Corollary 4.6.14 *Let H be a positive $\mathcal{F}$-measurable contingent claim. Then, under the assumptions of Theorem* 4.6.10, *for any utility function U, the utility indifference price coincides with the real-world price* (*at $t = 0$*), *namely,*

$$
p(H) = \mathbb{E}\left[\frac{H}{V_T^{\pi^*}}\right].
$$

Proof The present assumptions imply that, due to (4.28), we can rewrite (4.32) as follows:

$$
p(H) = \frac{\mathbb{E}[U'(I(\frac{\mathcal{Y}(v)}{\bar{V}_T^{v,\pi^*}}))\,\bar{H}]}{\mathbb{E}[U'(I(\frac{\mathcal{Y}(v)}{\bar{V}_T^{v,\pi^*}}))\,\bar{V}_T^{\pi^U}]} = \frac{\mathbb{E}[\frac{\mathcal{Y}(v)}{\bar{V}_T^{v,\pi^*}}\,\bar{H}]}{\mathbb{E}[\frac{\mathcal{Y}(v)}{\bar{V}_T^{v,\pi^*}}\,\bar{V}_T^{\pi^U}]} = \frac{\frac{1}{v}\,\mathbb{E}[\frac{\bar{H}}{\bar{V}_T^{\pi^*}}]}{\frac{1}{v}\frac{\bar{V}_0^{\pi^U}}{\bar{V}_0^{\pi^*}}} = \mathbb{E}\left[\frac{H}{V_T^{\pi^*}}\right], \tag{4.34}
$$

where the third equality uses the fact that $\pi^U \in \mathcal{A}^F$. □

Remark 4.6.15 As can be seen from Definition 4.6.12, the utility indifference price $p(H)$ depends a priori both on the initial endowment v and on the chosen utility function U. The remarkable result of Corollary 4.6.14 consists in the fact that, under the hypotheses of Theorem 4.6.10, the utility indifference price $p(H)$ represents a "universal" pricing rule, since it depends neither on v nor on the utility function U, and, furthermore, it coincides with the real-world pricing formula.

4.7 Conclusions, Extensions and Further Developments

In this work, we have studied a general class of diffusion-based models for financial markets, weakening the traditional assumption that the NFLVR condition holds or, equivalently, that there exists an ELMM. We have shown that the financial market may still be viable, in the sense that arbitrages of the first kind are not permitted, as soon as the market price of risk process satisfies a crucial square-integrability condition. In particular, we have shown that the failure of the existence of an ELMM does not preclude the completeness of the financial market and the solvability of portfolio optimisation problems. Furthermore, in the context of a complete market, contingent claims can be consistently evaluated by relying on the real-world pricing formula.

We have chosen to work in the context of a multi-dimensional diffusion-based modelling structure since this allows us to consider many popular and widely employed financial models and, at the same time, avoid some of the technicalities which arise in more general settings. However, most of the results of the present paper carry over to a more general and abstract setting based on continuous semimartingales, as shown in Chap. 4 of Fontana (2012). In particular, the latter work also deals with the robustness of the absence of arbitrages of the first kind with respect to several changes in the underlying modelling structure, namely changes of numéraire, absolutely continuous changes of the reference probability measure and restrictions and enlargements of the reference filtration.

The results of Sect. 4.6.3 on the valuation of contingent claims have been obtained under the assumption of a complete financial market. These results, in particular the fact that the real-world pricing formula (4.22) coincides with the utility indifference price, can be extended to the more general context of an incomplete financial market, provided that we choose a logarithmic utility function.

Proposition 4.7.1 *Suppose that Assumption* 4.3.7 *holds and let H be a positive $\mathcal{F}$-measurable contingent claim such that $\mathbb{E}[\widehat{Z}_T H/S_T^0] < \infty$. Then for $U(x) = \log(x)$, the log-utility indifference price $p_{\log}(H)$ is explicitly given as follows*:

$$p_{\log}(H) = \mathbb{E}\left[\frac{H}{V_T^{\pi^*}}\right].$$

Proof Note first that $U(x) = \log(x)$ is a well-defined utility function in the sense of Definition 4.6.8. Let us first consider problem (4.26) for $U(x) = \log(x)$. Using the notation introduced in the proof of Theorem 4.6.10, the function I is now given by $I(x) = x^{-1}$ for $x \in (0, \infty)$. Due to Proposition 4.4.4, we have $\mathcal{W}(y) = v/y$ for all $y \in (0, \infty)$ and, hence, $\mathcal{Y}(v) = 1$. Then, Eq. (4.28) directly implies that $\bar{V}_T^{v,\pi^U} = \bar{V}_T^{v,\pi^*}$, meaning that the growth-optimal strategy $\pi^* \in \mathcal{A}^F$ solves problem (4.26) for a logarithmic utility function. The same computations as in (4.34) imply then the following:

$$p_{\log}(H) = \frac{\mathbb{E}[\frac{\bar{H}}{\bar{V}_T^{v,\pi^*}}]}{\mathbb{E}[\frac{1}{\bar{V}_T^{v,\pi^*}}\bar{V}_T^{\pi^*}]} = \mathbb{E}\left[\frac{H}{V_T^{\pi^*}}\right]. \qquad \square$$

The interesting feature of Proposition 4.7.1 is that the claim H does not need to be replicable. However, Proposition 4.7.1 depends on the choice of the logarithmic utility function and does not hold for a generic utility function U, unlike the "universal" result shown in Corollary 4.6.14. Of course, the result of Proposition 4.7.1 is not surprising due to the well-known fact that the growth-optimal portfolio solves the log-utility maximisation problem, see e.g. Becherer (2001), Christensen and Larsen (2007) and Karatzas and Kardaras (2007).

Remark 4.7.2 Following Sect. 11.3 of Platen and Heath (2006), let us suppose that the discounted GOP process $\bar{V}^{\pi^*} = (\bar{V}_t^{\pi^*})_{0 \le t \le T}$ has the Markov property under P. Under this assumption, one can obtain an analogous version of Theorem 4.6.10 also in the case of an incomplete financial market model (see Platen and Heath 2006, Theorem 11.3.3). In fact, the first part of the proof of Theorem 4.6.10 remains unchanged. One then proceeds by considering the martingale $M = (M_t)_{0 \le t \le T}$ defined by $M_t := \mathbb{E}[\widehat{Z}_T\, I(\mathcal{Y}(v)/\bar{V}_T^{v,\pi^*})|\mathcal{F}_t] = \mathbb{E}[1/\bar{V}_T^{\pi^*}\, I(\mathcal{Y}(v)/\bar{V}_T^{v,\pi^*})|\mathcal{F}_t]$ for $t \in [0,T]$. Due to the Markov property, the martingale M_t can be represented as $g(t, \bar{V}_t^{\pi^*})$ for every $t \in [0,T]$. If the function g is sufficiently smooth one can apply Itô's formula and express M as the value process of a benchmarked fair portfolio. If one can show that the resulting strategy satisfies the admissibility conditions (see Definition 4.2.3), Proposition 4.6.13 and Corollary 4.6.14 can then be applied to show that the real-world pricing formula coincides with the utility indifference price (for any utility function!). Always in a diffusion-based Markovian context, a related analysis can also be found in the recent paper Ruf (2012).

We want to point out that the modelling framework considered in this work is not restricted to stock markets, but can also be applied to the valuation of fixed income products. In particular, in Bruti-Liberati et al. (2010) and Platen and Heath (2006), Sect. 10.4, the authors develop a version of the Heath–Jarrow–Morton approach to the modelling of the term structure of interest rates without relying on the existence of a martingale measure. In this context, they derive a real-world version of the classical Heath–Jarrow–Morton drift condition, relating the drift and diffusion terms in the system of SDEs describing the evolution of forward interest rates. Unlike in the traditional setting, this real-world drift condition explicitly involves the market price of risk process.

Finally, we want to mention that the concept of real-world pricing has also been studied in the context of incomplete information models, meaning that investors are supposed to have access only to the information contained in a sub-filtration of the original full-information filtration $\mathbb{F}$, see Galesso and Runggaldier (2010) and Platen and Runggaldier (2005, 2007).

Acknowledgements Part of this work has been inspired by a series of research seminars organised by the second author at the Department of Mathematics of the Ludwig-Maximilians-Universität München during the Fall Semester 2009. The first author gratefully acknowledges financial support from the "Nicola Bruti-Liberati" scholarship for studies in Quantitative Finance. We thank an anonymous referee for the careful reading and for several comments that contributed to improve the paper.

References

Ansel, J. P., & Stricker, C. (1992). Lois de martingale, densités et décomposition de Föllmer Schweizer. *Annales de l'Institut Henri Poincaré (B)*, *28*(3), 375–392.

Ansel, J. P., & Stricker, C. (1993a). Décomposition de Kunita-Watanabe. In *Séminaire de probabilités XXVII* (pp. 30–32).

Ansel, J. P., & Stricker, C. (1993b). Unicité et existence de la loi minimale. In *Séminaire de probabilités XXVII* (pp. 22–29).

Artzner, P., Delbaen, F., Eber, J. M., & Heath, D. (1999). Coherent measures of risk. *Mathematical Finance*, *9*, 203–228.

Ball, C. A., & Torous, W. N. (1983). Bond price dynamics and options. *Journal of Financial and Quantitative Analysis*, *18*(4), 517–531.

Becherer, D. (2001). The numeraire portfolio for unbounded semimartingales. *Finance and Stochastics*, *5*, 327–341.

Bernoulli, D. (1738). Specimen theoriae novae de mensura sortis. In *Commentarii Academiae Scientarium Imperialis Petropolitanae*.

Bernstein, P. L. (1996). *Against the gods: the remarkable story of risk*. New York: Wiley.

Bruti-Liberati, N., Nikitopoulos-Sklibosios, C., & Platen, E. (2010). Real-world jump-diffusion term structure models. *Quantitative Finance*, *10*(1), 23–37.

Cerreia-Vioglio, S. (2009). *Maxmin expected utility on a subjective state space: convex preferences under risk* (Preprint).

Cheridito, P., & Li, T. (2008). Dual characterization of properties of risk measures on Orlicz hearts. *Mathematics and Financial Economics*, *2*(1), 29–55.

Cheridito, P., & Li, T. (2009). Monetary risk measures on maximal subspaces of Orlicz classes. *Mathematical Finance*, *19*(2), 189–214.

Christensen, M. M., & Larsen, K. (2007). No arbitrage and the growth optimal portfolio. *Stochastic Analysis and Applications*, *25*, 255–280.

Cox, A. M. G., & Hobson, D. G. (2005). Local martingales, bubbles and options prices. *Finance and Stochastics*, *9*, 477–492.

Credit Suisse Financial Products (1997). *CreditRisk$^+$: a CreditRisk management framework*. London: Credit Suisse Financial Products.

Davis, M. H. A. (1997). Option pricing in incomplete markets. In M. A. H. Dempster & S. R. Pliska (Eds.), *Mathematics of derivative securities* (pp. 227–254). Cambridge: Cambridge University Press.

Delbaen, F., & Schachermayer, W. (1994). A general version of the fundamental theorem of asset pricing. *Mathematische Annalen*, *300*, 463–520.

Delbaen, F., & Schachermayer, W. (1995a). Arbitrage possibilities in Bessel processes and their relations to local martingales. *Probability Theory and Related Fields*, *102*, 357–366.

F. Biagini et al. (eds.), *Risk Measures and Attitudes*, EAA Series,
DOI 10.1007/978-1-4471-4926-2, © Springer-Verlag London 2013

Delbaen, F., & Schachermayer, W. (1995b). The existence of absolutely continuous local martingale measures. *Annals of Applied Probability*, *5*(4), 926–945.

Delbaen, F., & Schachermayer, W. (2006). *The mathematics of arbitrage*. Berlin: Springer.

Delbaen, F., Drapeau, S., & Kupper, M. (2011). A von Neumann–Morgenstern representation result without weak continuity assumption. *Journal of Mathematical Economics*, *47*, 401–408.

Denuit, M., & Eeckhoudt, L. (2010). Bivariate stochastic dominance and substitute risk (in)dependent utilities. *Decision Analysis*, *7*(3), 302–312.

Denuit, M., & Mesfioui, M. (2010). Generalized increasing convex and directionally convex orders. *Journal of Applied Probability*, *47*(1), 264–276.

Denuit, M., Lefèvre, C., & Mesfioui, M. (1999). A class of bivariate stochastic orderings, with applications in actuarial sciences. *Insurance. Mathematics & Economics*, *24*(1), 31–50.

Denuit, M., Eeckhoudt, L., & Rey, B. (2010). Some consequences of correlation aversion in decision science. *Annals of Operations Research*, *176*(1), 259–269.

Drapeau, S., & Kupper, M. (2010, forthcoming). Risk preferences and their robust representation. *Mathematics of Operations Research*.

Dunkel, J., & Weber, S. (2007). Efficient Monte Carlo methods for convex risk measures in portfolio credit risk models. In S. G. Henderson, B. Biller, M.-H. Hsieh, J. Shortle, J. D. Tew, & R. R. Barton (Eds.), *Proceedings of the 2007 winter simulation conference*, Washington, DC (pp. 958–966). Piscataway: IEEE.

Dunkel, J., & Weber, S. (2010). Stochastic root finding and efficient estimation of convex risk measures. *Operations Research*, *58*(5), 1505–1521.

Eeckhoudt, L., & Schlesinger, H. (2006). Putting risk in its proper place. *The American Economic Review*, *96*(1), 280–289.

Eeckhoudt, L., Rey, B., & Schlesinger, H. (2007). A good sign for multivariate risk taking. *Management Science*, *53*(1), 117–124.

Eeckhoudt, L., Schlesinger, H., & Tsetlin, I. (2009). Apportioning of risks via stochastic dominance. *Journal of Economic Theory*, *144*(3), 994–1003.

Ekern, S. (1980). Increasing *n*th degree risk. *Economics Letters*, *6*(4), 329–333.

Embrechts, P., McNeil, A. J., & Straumann, D. (2002). Correlation and dependency in risk management: properties and pitfalls. In M. Dempster (Ed.), *Risk management: value at risk and beyond* (pp. 176–223). Cambridge: Cambridge University Press.

Epstein, L. G., & Tanny, S. M. (1980). Increasing general correlation: a definition and some economic consequences. *Canadian Journal of Economics*, *13*(1), 16–34.

Fernholz, R., & Karatzas, I. (2009). Stochastic portfolio theory: an overview. In A. Bensoussan & Q. Zhang (Eds.), *Handbook of numerical analysis: Vol. XV. Mathematical Modeling and Numerical Methods in Finance* (pp. 89–167). Oxford: North-Holland.

Fishburn, P. C. (1974). Convex stochastic dominance with continuous distribution functions. *Journal of Economic Theory*, *7*(2), 143–158.

Fishburn, P. C. (1978). Convex stochastic dominance. In G. A. Whitmore & M. C. Findlay (Eds.), *Stochastic dominance: an approach to decision-making under risk* (pp. 337–351). Lexington: Heath.

Fishburn, P. C. (1982). *The foundations of expected utility*. Dordrecht: D. Reidel Publishing.

Föllmer, H., & Schied, A. (2002). Robust representation of convex measures of risk. In *Advances in finance and stochastics. Essays in honour of Dieter Sondermann* (pp. 39–56). Berlin: Springer.

Föllmer, H., & Schied, A. (2004). *de Gruyter studies in mathematics. Stochastic finance—an introduction in discrete time* (2nd ed.). Berlin: Walter de Gruyter.

Föllmer, H., & Schied, A. (2011). *Stochastic finance—an introduction in discrete time* (3rd ed.). Berlin: Walter de Gruyter.

Fontana, C. (2012). *Four essays in financial mathematics*. Ph.D. thesis, University of Padova.

Frey, R., & McNeil, A. J. (2002). VaR und expected shortfall in portfolios of dependent credit risks: conceptual and practical insights. *Journal of Banking & Finance*, *26*, 1317–1334.

Frey, R., & McNeil, A. J. (2003). Dependant defaults in models of portfolio credit risk. *The Journal of Risk*, *6*(1), 59–92.

Frey, R., Popp, M., & Weber, S. (2008). An approximation for credit portfolio losses. *The Journal of Credit Risk*, *4*(1), 3–20.

Galesso, G., & Runggaldier, W. (2010). Pricing without equivalent martingale measures under complete and incomplete observation. In C. Chiarella & A. Novikov (Eds.), *Contemporary quantitative finance: essays in honour of Eckhard Platen* (pp. 99–121). Berlin: Springer.

Giesecke, K., Schmidt, T., & Weber, S. (2008). Measuring the risk of large losses. *Journal of Investment Management*, *6*(4), 1–15.

Glasserman, P. (2004). *Applications of mathematics: Vol. 53. Monte Carlo methods in financial engineering*. New York: Springer.

Glasserman, P., Heidelberger, P., & Shahabuddin, P. (2002). Portfolio value-at-risk with heavy-tailed risk factors. *Mathematical Finance*, *12*(3), 239–269.

Gordy, M. (2000). A comparative anatomy of credit risk models. *Journal of Banking & Finance*, *24*, 119–149.

Gupton, C., Finger, C., & Bhatia, M. (1997). *CreditMetrics technical document*. New York: J. P. Morgan & Co. www.riskmetrics.com.

Hadar, J., & Russell, W. R. (1969). Rules for ordering uncertain prospects. *The American Economic Review*, *59*(1), 25–34.

Hanoch, G., & Levy, H. (1969). The efficiency analysis of choices involving risk. *Review of Economic Studies*, *36*(3), 335–346.

Hazen, G. B. (1986). Partial information, dominance, and potential optimality in multiattribute utility theory. *Operations Research*, *34*(2), 296–310.

Heston, S. L., Loewenstein, M., & Willard, G. A. (2007). Options and bubbles. *The Review of Financial Studies*, *20*(2), 359–390.

Hulley, H. (2010). The economic plausibility of strict local martingales in financial modelling. In C. Chiarella & A. Novikov (Eds.), *Contemporary quantitative finance: essays in honour of Eckhard Platen* (pp. 53–75). Berlin: Springer.

Hulley, H., & Schweizer, M. (2010). M^6—on minimal market models and minimal martingale measures. In C. Chiarella & A. Novikov (Eds.), *Contemporary quantitative finance: essays in honour of Eckhard Platen* (pp. 35–51). Berlin: Springer.

Jarrow, R. A., & Madan, D. B. (1999). Hedging contingent claims on semimartingales. *Finance and Stochastics*, *3*, 111–134.

Jarrow, R. A., Protter, P., & Shimbo, K. (2007). Asset price bubbles in complete markets. In M. C. Fu, R. A. Jarrow, J. Y. J. Yen, & R. J. Elliott (Eds.), *Advances in mathematical finance* (pp. 97–122). Boston: Birkhäuser.

Jarrow, R. A., Protter, P., & Shimbo, K. (2010). Asset price bubbles in incomplete markets. *Mathematical Finance*, *20*(2), 145–185.

Jorion, P. (2000). *Value at risk* (2nd ed.). New York: McGraw-Hill.

Kabanov, Y. (1997). On the ftap of Kreps-Delbaen-Schachermayer. In Y. Kabanov, B. L. Rozovskii, & A. N. Shiryaev (Eds.), *Statistics and control of stochastic processes: the Liptser festschrift* (pp. 191–203). Singapore: World Scientific.

Kang, W., & Shahabuddin, P. (2005). Fast simulation for multifactor portfolio credit risk in the t-copula model. In M. E. Kuhl, N. M. Steiger, F. B. Armstrong, & J. A. Joines (Eds.), *Proceedings of the 2005 winter simulation conference* (pp. 1859–1868). Hanover: INFORMS.

Karatzas, I., & Kardaras, K. (2007). The numeraire portfolio in semimartingale financial models. *Finance and Stochastics*, *11*, 447–493.

Karatzas, I., & Shreve, S. E. (1991). *Brownian motion and stochastic calculus* (2nd ed.). New York: Springer.

Karatzas, I., & Shreve, S. E. (1998). *Methods of mathematical finance*. New York: Springer.

Kardaras, C. (2010). Finitely additive probabilities and the fundamental theorem of asset pricing. In C. Chiarella & A. Novikov (Eds.), *Contemporary quantitative finance: essays in honour of Eckhard Platen* (pp. 19–34). Berlin: Springer.

Kardaras, C. (2012). Market viability via absence of arbitrage of the first kind. *Finance and Stochastics*, *16*, 651–667.

Keeney, R. L. (1992). *Value-focused thinking*. Cambridge: Harvard University Press.

Keeney, R., & Raiffa, H. (1976). *Decisions with multiple objectives: preferences and value trade-offs*. New York: Wiley.

Levental, S., & Skorohod, A. S. (1995). A necessary and sufficient condition for absence of arbitrage with tame portfolios. *Annals of Applied Probability*, *5*(4), 906–925.

Levhari, D., Paroush, J., & Peleg, B. (1975). Efficiency analysis of multivariate distributions. *Review of Economic Studies*, *42*(1), 87–91.

Levy, H., & Paroush, J. (1974). Toward multi-variate efficiency criteria. *Journal of Economic Theory*, *7*(2), 129–142.

Loewenstein, M., & Willard, G. A. (2000). Local martingales, arbitrage, and viability: free snacks and cheap thrills. *Economic Theory*, *16*(1), 135–161.

Londono, J. A. (2004). State tameness: a new approach for credit constraints. *Electronic Communications in Probability*, *9*, 1–13.

MacLean, L. C., Thorp, E. O., & Ziemba, W. T. (2010). Long-term capital growth: the good and bad properties of the Kelly and fractional Kelly capital growth criteria. *Quantitative Finance*, *10*(7), 681–687.

McNeil, A. J., Frey, R., & Embrechts, P. (2005). *Princeton series in finance. Quantitative risk management: concepts, techniques and tools*. Princeton: Princeton University Press.

Menezes, C., Geiss, C., & Tressler, J. (1980). Increasing downside risk. *The American Economic Review*, *70*(5), 921–932.

Mosler, K. C. (1984). Stochastic dominance decision rules when the attributes are utility independent. *Management Science*, *30*(11), 1311–1322.

Müller, A. (2001). Stochastic ordering of multivariate normal distributions. *Annals of the Institute of Statistical Mathematics*, *53*(3), 567–575.

Müller, A., & Stoyan, D. (2002). *Comparison methods for stochastic models and risks*. Chichester: Wiley.

Platen, E. (2002). Arbitrage in continuous complete markets. *Advances in Applied Probability*, *34*, 540–558.

Platen, E. (2006). A benchmark approach to finance. *Mathematical Finance*, *16*(1), 131–151.

Platen, E. (2011). A benchmark approach to investing and pricing. In L. C. MacLean, E. O. Thorp, & W. T. Ziemba (Eds.), *The Kelly capital growth investment criterion* (pp. 409–427). Singapore: World Scientific.

Platen, E., & Heath, D. (2006). *A benchmark approach to quantitative finance*. Berlin: Springer.

Platen, E., & Runggaldier, W. J. (2005). A benchmark approach to filtering in finance. *Asia-Pacific Financial Markets*, *11*, 79–105.

Platen, E., & Runggaldier, W. J. (2007). A benchmark approach to portfolio optimization under partial information. *Asia-Pacific Financial Markets*, *14*, 25–43.

Polyak, B. T., & Juditsky, A. B. (1992). Acceleration of stochastic approximation by averaging. *SIAM Journal on Control and Optimization*, *30*, 838–855.

Press, W. H., Vetterling, W. T., & Flannery, B. P. (2002). *Numerical recipes in C++: the art of scientific computing* (2nd ed.). Cambridge: Cambridge University Press.

Revuz, D., & Yor, M. (1999). *Continuous martingales and Brownian motion* (3rd ed.). Berlin: Springer.

Richard, S. F. (1975). Multivariate risk aversion, utility independence and separable utility functions. *Management Science*, *22*(1), 12–21.

Rothschild, M., & Stiglitz, J. E. (1970). Increasing risk: I. A definition. *Journal of Economic Theory*, *2*(3), 225–243.

Ruf, J. (2012). Hedging under arbitrage. *Mathematical Finance*, to appear.

Ruppert, D. (1988). *Efficient estimators from a slowly convergent Robbins-Monro procedure* (ORIE Technical Report 781). Cornell University.

Ruppert, D. (1991). Stochastic approximation. In B. Gosh & P. Sen (Eds.), *Handbook of sequential analysis* (pp. 503–529). New York: Marcel Dekker.

Scarsini, M. (1988). Dominance conditions for multivariate utility functions. *Management Science*, *34*(4), 454–460.

Schweizer, M. (1992). Martingale densities for general asset prices. *Journal of Mathematical Economics*, *21*, 363–378.

Schweizer, M. (1995). On the minimal martingale measure and the Fllmer-Schweizer decomposition. *Stochastic Analysis and Applications*, *13*, 573–599.

Shaked, M., & Shantikumar, J. G. (2007). *Stochastic orders*. New York: Springer.

Strasser, E. (2005). Characterization of arbitrage-free markets. *Annals of Applied Probability*, *15*(1A), 116–124.

Tasche, D. (2002). Expected shortfall and beyond. *Journal of Banking & Finance*, *26*(7), 1519–1533.

Tsetlin, I., & Winkler, R. L. (2005). Risky choices and correlated background risk. *Management Science*, *51*(9), 1336–1345.

Tsetlin, I., & Winkler, R. L. (2007). Decision making with multiattribute performance targets: the impact of changes in performance and target distributions. *Operations Research*, *55*(2), 226–233.

Tsetlin, I., & Winkler, R. L. (2009). Multiattribute utility satisfying a preference for combining good with bad. *Management Science*, *55*(12), 1942–1952.

von Neumann, J., & Morgenstern, O. (1947). *Theory of games and economics behavior* (2nd ed.). Princeton: Princeton University Press.

Weber, S. (2006). Distribution-invariant risk measures, information, and dynamic consistency. *Mathematical Finance*, *16*(2), 419–442.

Index

F. Biagini et al. (eds.), *Risk Measures and Attitudes*, EAA Series,
DOI 10.1007/978-1-4471-4926-2, © Springer-Verlag London 2013

MIX
Papier aus verantwortungsvollen Quellen
Paper from responsible sources
FSC® C105338

If you have any concerns about our products,
you can contact us on
ProductSafety@springernature.com

In case Publisher is established outside the EU,
the EU authorized representative is:
Springer Nature Customer Service Center GmbH
Europaplatz 3, 69115 Heidelberg, Germany

Printed by Libri Plureos GmbH
in Hamburg, Germany